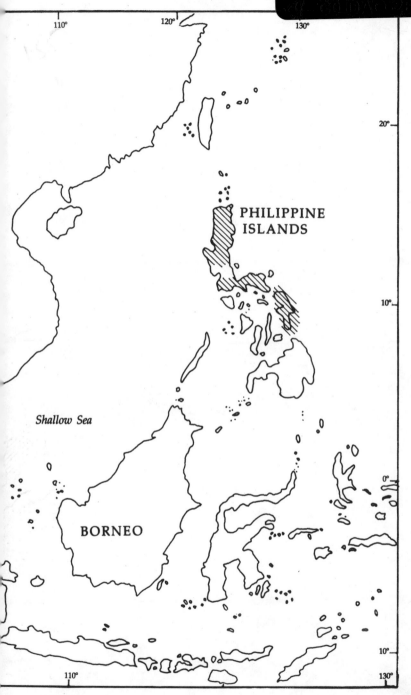

110°   120°   130°

20°

PHILIPPINE
ISLANDS

10°

Shallow Sea

0°

BORNEO

10°

110°   130°

ABOVE THE FOREST

ABOVE THE FOREST
A Study of Andamanese Ethnoanemology, Cosmology, and the Power of Ritual

Vishvajit Pandya

DELHI
OXFORD UNIVERSITY PRESS
BOMBAY CALCUTTA MADRAS
1993

Oxford University Press, Walton Street, Oxford OX2 6DP

New York Toronto
Delhi Bombay Calcutta Madras Karachi
Kuala Lumpur Singapore Hong Kong Tokyo
Nairobi Dar es Salaam
Melbourne Auckland
and associates in
Berlin Ibadan

ISBN 0 19 562971 X

Typeset by Resodyn, New Delhi 110030
Printed in India at Rekha Printers Pvt. Ltd., New Delhi 110020
and published by Neil O'Brien, Oxford University Press
YMCA Library Building, Jai Singh Road, New Delhi 110001

For
all the Ongees who shared
my parents who supported
and
Atayih who is curious

Preface

It could be argued that, with the exception of a doctrine of cultural relativism, anthropology's contribution to Western thought and letters has been methodological rather than theoretical. Anthropology is certainly exceptional among the Social Sciences for its commitment to lengthy periods of qualitative field research and its techniques of participant observation. As uncodified and irreproducable as they are, anthropology produces a unique way of knowing. As the product of fieldwork, ethnography has been the discipline's principal means of communicating it's unique way of knowing. To go and live with people very different from one's own, to accumulate some experience and write about it, is the tradition of anthropology. What is studied is based on questions, sometimes about the culture studied and sometimes based on the culture within which the investigator is contextualized. Therefore, the anthropologist either takes a text to a context or brings a context to a text. Originally my study of the Andaman Islands was based on the idea of taking the text created by Radcliffe-Brown to its context, and to create another text based on the supposition that the context may have changed, and the texts within the anthropological tradition have certainly changed since the first publication of Radcliffe-Brown's *Andaman Islanders* in 1922 and revised edition in 1964. Durkheim's *Elementary Forms of Religious Life*, and C. Lévi-Strauss's *The Elementary Structure of Kinship* inspired me to study something 'elementary' and 'structural', but ritual was my prime concern. Perhaps this was based on my upbringing within a moderately conservative household much obsessed with values of 'Pure' and 'Impure'. This was the orientation with which I went to the field, with a proposition to study the efficacy and power of ritual. Knowing

that ritual has power, the question was where did the power
come from, how was it procured and used in as elementary a
structure of ritual as that amongst the Andaman Islanders, and
their belief system, which has an essential unity of 'nature'
(including humans) and 'supernature' (including supernatural
beings, i.e. spirits).

Indeed the so-called study of 'ritual' has included under it
various situations of public, ritualized, and sanctioned exer-
cises of power, where invocation, possession, curing, and transaction, recitation, and act are all included as what was
performed and what was observed. This leads to analysis of
various ritual elements in one or more analytic moulds—providing a key to significant aspects of the sociocultural order. In
such attempts and approaches, ritual becomes a conglomerate
of various elements that are used to build a scheme that explicates the everyday order.

On reaching the forest home of the Ongees, a question that
was continuously on my mind was: When will this camp move
from here to another place? It was the month of October 1983,
and I had been living at the coastal camp near Dugong Creek
for a month. I was in a place where, to our east was the
Andaman sea, and on the west was the deep tropical rain forest
with enormous trees, in many cases reaching over a hundred
feet in height. The tree-tops form a canopy that cuts out all
direct sunlight on the ground. Although the trees are so large
and heavy, their roots tend to be shallow and they are easily
toppled over in the heavy storms that frequent the Little Andamans. The Ongees were aware of all this, so they were
waiting for the wind conditions to change before moving
into the forest. I was tired of eating fish and turtle meat and
wondered when I would be moving with the Ongees into the
thick tropical forest, which was so overwhelming and appealing from the white sands of Dugong Creek. For the Ongees,
living on the coast was a safe place, but for me the coastal camp
was just one place among other places that I wanted to visit
with the Ongees. My city background made me eager to enter
the forest, but a sense of fear overpowered curiosity and kept
me tied to the then strange camp-mates with whom I was
living. To move from place to place in a given space for me was
the essence of freedom. The associations of place with safety

and security, and of space with freedom and exploration, were my first realizations about our culture, while living with the Ongees at Dugong Creek.

In due course I learned the *ceye?ne* language of the Ongees. In *ceye?ne* the use of prefixes and suffixes makes it possible to trace meanings and ideas in each word (Bloch, 1949; Radcliffe-Brown, 1914). For example, *ko* means 'where' and *ale* means 'children'. By putting together *ko-r-ale* the term becomes *korale*, 'home, a place where children are'. As I started acquiring a better grasp of the language, I started posing the question to all the people around me: When will we move from here to another place? I never got any specific answer in terms of time. There was only one answer, and it was rhetorical: 'This place is good when the forest is good we move into forest place—then the sea place is bad!'

What makes a place good and bad or safe and dangerous, and when do settlement and movement make sense? This was the first question in my notebook. Besides, I also had to find out if the concerns in my mind were questions for the Ongee in his own cultural context. It is important to understand how different cultures would and do answer questions. However, the questions remain questions only for anthropologists; are they also questions for the people in a given culture? This account of Ongee culture is structured around the questions that are important for the Andamanese as well as significant within Andamanese ethnography.

I learned that how and when the Ongees move depend on all that moves within the Ongee space. Just as I was interested in mapping out the movement of the Ongee hunters and gatherers, the islanders were on the look-out for what moved around them and what they could do in relation to the various aspects of movement and to the elements that moved. For the Ongees, the elements referred not only to the natural aspects of Ongee space, such as the tides, sun, moon, clouds, earthquakes, and winds, but also to the animals, such as pigs, turtles, dugongs, monitor lizards, civet cats, and insects, as well as to the spirits. These elements constitute an integral part of the Ongee world-view. The places in which the elements move are interrelated by the relation of act and impact within the whole space. Every element's movement, an act, has an impact on other elements. This combination of the elements' act–impact relation sets up a series of transformations. This viewpoint makes the Ongee cosmology a stage of movements where space and place are never empty. The awareness of different elements and their movements within different places, guides Ongee action. Consequently, the cosmology in which Ongees, elements, places, and movement are located acquires an ideological value. The ideological value and the cartographical implications of Ongee cosmology organizes the world of the Ongees in which social processes are effective. The Ongee cosmology, however, is not based on just the idea of nature, but also on how nature is related to the social. The recognition of this relationship between nature, where things move, and society, where things are accordingly moved, leads to the emergence of various values as cultural reality. Equality and hierarchy of humans and spirits, things and ideas, are therefore imbedded in the context of Ongee cosmology, and are signifiers and signified through the various forms of movement of various elements. Spirits, human beings, and animals are distinguished from one another on the basis of their differing moving capacities and body conditions. They are in a hierarchical relationship with one another in which a body that has smell and a body that does not have smell are 'encompassed' elements. Bodies that move on both the vertical and horizontal axes and bodies that move only on one axis or the other are the 'encompassing' features of the hierarchical system.

To understand Ongee cosmology as a 'structure', in the sense that L. Dumont (1970) uses it, the fundamental opposition between humans and spirits and between humans and animals must be located. This opposition is based on the possession of smell: bodies without smell move along both the vertical and horizontal axes of Ongee space. The loss of smell or the acquisition of smell changes the position of bodies within the hierarchical structure. The whole of Ongee cosmology is founded on the necessary and hierarchical coexistence of spirits, humans, and animals. This conceptualized structural universe is important in Dumont's understanding of the Indian caste system (Dumont, 1970: 43–4); a system in which the whole governs the parts, and this whole is conceived to be based on an opposition. The Ongees and other related tribes of Andaman

Islanders do not exhibit the intricate social stratification evident in South Asian ethnography. However, Dumont's method
of imposing an opposition based on the whole that governs the
parts is applicable to the Ongee cosmological structure and the
cultural system. Dumont's method is particularly useful in
understanding certain aspects within Ongee culture such as
why, within the system of hierarchy, spirits, humans, and
animals are subject to changing positions. Within the Ongee
system of hierarchy, that which is 'encompassed' becomes
'encompassing'. This is what characterizes Ongee cosmology
and hierarchy and is the point at which my explanation and
limitation of Dumont's construct of hierarchical structure diverges.

The Ongees believe that the death of human beings, loss of
weight, and the dispersal of smell are all ways and steps in
which humans are transformed into spirits. It is these spirits
who then can move any and everywhere. Ongees also believe
that although spirits, humans, and animals share a common
space, it is through their different capacities of movement that
each remains alive within different places within that space.
Spirits hunt humans and displace them by killing and taking
them away from the island. Human beings, like the spirits, hunt
animals and take them away from either the sea or forest. This
relationship of hunting, loss of life, and displacement makes it
necessary to regard the hierarchical relations within the 'system' as not just 'apparently static' but 'consistently dynamic'.
Leach (1967) suggests that the structures which the anthropologist describes are the anthropological conception of structure reflecting a 'real society', which 'is a process in time'
(Leach, 1967: 5), the factors that effect movements, smell, and
consequent changes in the hierarchical structure are important
in this presentation of ethnography. The Ongees' relationship
with spirits and animals is not to be viewed as only an 'as if'
model (Leach, 1967), but also must consider a system of relations that actually do exist for the Ongees. For the Ongees the
system of relations within the structure changes and is based
on the concept of controlling the movement of smell in order
to effect the movement of bodies that in turn brings about
changes in relations. Therefore, I intend to go beyond the level
of abstraction typically used by the anthropologist when

describing a social structure solely in terms of the principles of organization that unite the component parts of the system and by which structure exists independent of cultural content (cf. Fortes, 1949: 54–60). I intend to present the cultural content and concepts of cosmological structures and hierarchical systems. This conception of structure does not exist only for the anthropologist or in the anthropologist's mind and is not simply logically created and derived or influenced by, say, the African pastoralist or South American native. Rather, it is a structure conceived on the basis of concepts perceived by the Ongees themselves.

Ongees believe that spirits transform human beings into food or fellow spirits; human beings transform spirits into fetuses and transform animals from raw meat to cooked food. These are all concrete cases of changes in the position of individuals within the 'anthropologist's system' as well as the 'system' that is real for the Ongees. Changes of position are caused through *enakyu?la* ('power) that makes it possible for elements and individuals within the hierarchical system to be placed, displaced, and replaced within space. *Enakyu?la* also effects the interaction and interactional outcome of living as a hunter and gatherer. The Ongees describe *enakyu?la* as the ability to displace or replace an element. For example, a person who pushes a canoe from the beach and into the sea is believed to be using *enakyu?la*. Similarly, the spirit who takes a human being up to the sky and the Ongee hunter who brings his kill back to the camp-site demonstrates *enakyu?la*. These examples of *enakyu?la* involve movement during which an object is transformed. Changes in location also reflect a change in hierarchical position.

Within the Ongee world-view, power is not only related to aspects of change but also to changes in power relations implicit in the hierarchy. Thus, *enakyu?la* pertains both to location and movement attributed to each element that occupies a position in a place within the space. The position of an element and its capacity to move are the attached aspects of power. Each element within this structure wields power in terms of its capacity to move and to replace or displace other elements in the hierarchy. All Ongee actions involving the exercise of *enakyu?la* are consciously directed towards a particular end.

The Ongee's concern with power, and my depiction of that concern, is in no way a power argument concerning 'needs' and 'goals', as in the works of Malinowski (1944) and Parsons (1949, 1951). As an ethnographer my goal is to present the Ongees' concern with gaining power as a motive in society. An Ongee faces a choice of action, that is, either to have a *talabuka* ('conjunction') to gain power or a *malabuka* ('coincidence) to avoid experiencing power exerted over them. These choices, *talabuka* and *malabuka*, within Ongee culture are articulated and symbolized through *kwayabe* ('smell'). The dispersal and retention of smell defines a series of actual, real, past, present, and future relations.

My use of the term 'symbol' is not restricted to 'something that stands for something else', or something else 'where there is no necessary or intrinsic relationship between the symbol and that which it symbolizes' (Schneider, 1968). The symbols and the meaning of those symbols constitute a system because the people in a given culture share these meanings and connect them to one another. How people connect symbols and meanings is important since symbols are often thought of as 'things'. The concept of symbol, as I use it, also includes actions. Some symbolic acts are more central than others to a particular culture. Such central symbolic acts are typically laden with a greater than usual variety of meanings, that is, they are 'multivocal' or 'polysemic' (Turner, 1967, 1969). The central symbolic actions within Ongee culture, *talabuka* and *malabuka*, are not only symbolizing but also actions that, when properly done, achieve the objective of coincidence or conjunction through which the power positions of spirits, humans, and animals, based on the place attributed to them in cosmology and hierarchy, is maintained. The symbolic actions of *malabuka* and *talabuka* pertain to smell. Ideas related to smell are the basis of the total Ongee conceptualization of their cosmological structure as well as power relations. Within this conceptualization one can trace total information enabling 'focalization' and 'evocation' (Sperber, 1975: 119) of 'structure of practice' and 'practice of structure' (Sahlins, 1980). The central core around which power relations are established and re-established is the symbolism of smell. The centrality of smell and its movement in Ongee culture is similar to Stanley Tambiah's (1976) notion of

galactic polity in which the king is like a powerful sun-like
object. The king's gravitational pull keeps in orbit, at some
distance from itself, an unspecified number of lesser rulers,
each a simulation of the leading king. The whole is unified by
a field force characterized by both repulsion and attraction. In
galactic polity, power, like light, is in the centre of the system,
and it loses strength as it reaches the periphery. The symbolism
of centrality of power posits that there is a system even when
one cannot locate a specific power 'mechanism'. The centrality
of galactic polity characterizes the description of power and
state in South and South-East Asian culture as having ritual
sovereignty. Ritual sovereignty includes symbols and processes that, in the absence of instrumental mechanisms such as
taxes and coercion and checks and law within a state, creates a
domain or a realm of power. A classic example of this is in
Clifford Geertz's *Negara* (1980) that focuses on the ceremonial
and theatrical creation of the monarch as well as god (especially
the aspect of the king's *digvijayan* movement) (cf. Inden, 1986).
For Geertz, the state, a form of power, is constructed by the king
and the king is constructed by constructing a god (Geertz, 1980:
124). The king, an incarnation of the holy, is created through
ritual. The state draws its power from the symbolic capacity to
enchant though the king remains a symbol of divinity in the
cosmological centre. The king in other descriptions of South
and South-East Asian societies is an upholder of the social
order and fits into cosmology. Spatial cosmoses are often constructed through courts and temples that connect the order of
the person, the society, the king, and the world to the gods, as
in Hinduism, or to a transcendent state of being, as in Buddhism. These are cosmologies that construe a ruler as mediating between the forces of 'here' and 'there'.

   Returning to the Ongee notion of power, its symbolism is
central but there is no distinction between the powerful (king
and god) and the powerless (subject of the king). The mechanism of power replicates for and is accessible to all. Therefore, there are neither rituals dealing specifically with power
nor are there specific individuals who have the power of conducting rituals. Ongee actions pertaining to power go beyond
the Durkheimian (1915) distinction of actions into major classes, namely, religious rites that are sacred and technical acts

that are profane. Without addressing the controversy between the way Malinowski's placement of magic as sacred (Malinowski, 1948: 67) and Mauss regarding it as profane (Mauss, 1947: 207), my experience with the Ongees makes me question the assumption that the sacred (religious) and the profane (technical) are distinct wholes. Central symbolism and symbolic acts pertaining to smell and power in Ongee society makes the religious and technical, sacred and profane, a continuum and not separate categories (cf. Endicott, 1970, 1977). The central symbolism of smell destroys the anthropological categorization of ritual and secular.

The dynamic aspect between the hunter and the hunted, as embodied in the central symbols of smell and power, escaped Radcliffe-Brown's (1922) account of the Andaman Islanders. Influenced by Harrison (1912, 1913), Radcliffe-Brown in *Andaman Islanders* ([1922] 1964) developed the concept of 'ritual value' attached to certain objects that are socially important for secular reasons. From Radcliffe-Brown's perspective, the performance of ritual generates in the actors certain 'sentiments' that are advantageous to the society as a whole. Within this framework, Radcliffe-Brown interpreted ritual and postulated that human beings always manipulate their thought categories in a consistent way. The meaning of ritual symbols could be discovered by observing the diverse use of those symbols in both ritual and secular contexts. This was perhaps a good methodological idea, but not honest to the ethnographic reality. Radcliffe-Brown's failure to comprehend the manipulative system of thought categories consistently leads him to say, in reference to the question he posed to the Islanders about what becomes of man's spirit after death, that all their responses were 'different and inconsistent' (Radcliffe-Brown, 1964: 168). He believed that the Andamanese ideas on the subject to be 'floating and lacking in precision' (ibid.). I regard the 'inconsistency', 'floating', and lack of 'precision' to be that which characterizes the dynamic cosmological and hierarchical structure of the Islanders since there is no one place to which all dead men's spirits go. There is no 'here' and 'there' for humans and spirits. Humans and spirits coexist in a shared space, they have relations of power between them of such a character that humans can become spirits and spirits can become humans.

This dynamic was the reason why Radcliffe-Brown's informants give seemingly conflicting answers.

Radcliffe-Brown's informants in 1906 were from different
dialect groups. Under the British administration they had been
brought to reside at Port Blair in an effort to control the spread
of social diseases. Unlike Radcliffe-Brown, my fieldwork was
carried out on one island among only the Ongees. However,
the Ongees also gave various responses when I posed
Radcliffe-Brown's question about what happens to man after
death. The Ongees would never disagree with what other
Ongees had said about where the spirit of the dead person was.
Often one Ongee would incorporate another Ongee's idea and
further develop it. During my fieldwork it became clear to me
that only segments of an idea might be known yet all the
segments could be put together. The one constant in the explanations was the sequence of coming in and going out of
spirits in different places within the space. The response of the
Andamanese to Radcliffe-Brown's question indicates that for
them the *toma* ('spirits' (*tomya* in the Ongee language)) are not
located in one position and at one place. Spirits constantly
move around, depending on the relative movements of smells,
winds, and the Ongees themselves. Since the hierarchy of the
hunter and the hunted is subject to shifts and switches, the
dynamic nature of the structure in which relations are systematized is subject to change. Since this is the structure that
exists for the Ongee, the Ongee expresses it similarly. No
ethnographer should expect the Andamanese to say, 'This is
the earth, there is heaven, and that is hell'. Structure and
organization based in hierarchy and relations of power are
subject to constant changes. The Andamanese response is not
common in the great religions or the religions of the Book. As
R. Firth, in his 1955 Frazer lecture, says, 'primitive beliefs about
the fate of the soul are usually not polarized, as they are in the
great religions . . . most primitive eschatology is dynamic, with
plenty of social interaction' (Firth, 1967: 332).

To convey the dynamic nature of structure and the various
relations in it and the ways in which central symbols and
actions systematize meanings and concepts for Ongees, I have
focused on the issue of smell and power. First I 'define' the
'phenomena' (Lévi-Strauss, 1973: 84) of space and cosmology

under study' as a relation between two or more 'real or supposed' aspects such as what moves and where it moves, specifically concerning hunter and hunted, smells, and winds. Finally, I consider *malabuka* and *talabuka* as the general object of analysis to derive necessary connections to understand what *enakyu?la* is in empirical phenomenon such as the rituals of the spirit communicator and the initiation of the young men. In the ritual contexts, the multiple combinations of elements and aspects of smell and movement are used to formulate various relations involving spirits and humans in the cosmological structure and hierarchical system. Cosmology thus represents an aspect of culture in which structure and process meld, revealing cosmosophy. By cosmosophy I mean the philosophical basis and implications of cosmological structure in relation to social structure. Ongee cosmosophy is based on the allocation of power in cosmology—power that is neither evil nor good, but can produce either negative or positive results. Power, once acquired from the spirits, can be used either to counteract or create danger. Cosmosophy thus allows islanders with an orientation by means of which they can either acquire power, or take protective measures against it; both acquisition of power and protection from it affect Ongee movement and their perceptions of the space in which they move.

The time of my fieldwork went by as the Ongees moved from place to place on the island of Little Andaman. I realized that for the Ongees, movements through space, in relation to other things that moved, created the abstract as well as concrete idea of places being different. Instead of saying it is time to move, Ongees would emphasize the place into which they were moving. For Ongees time does not change, but various movements mark a change of places within space. Therefore, the time to move from the coast to the forest is when the coast becomes a bad place and the forest becomes a good place. The forest becomes a good place because it is there that the men and women moving around find food and will not be hunted and captured by the spirits. When men and women move into the forest, the spirits and winds move to the sea and coastal area, making the movement of the Ongees in that area difficult. This kind of explanation is perhaps difficult to translate and transcribe without using our idea of time and the terms pertaining to it. It is my intention to show how the Ongees relate space to place, spirits to man, nature to the social, with certain underlying principles through which hierarchy, equality, safety, and danger become temporally specified happenings in our perception, but for the Ongees remain movements in place through which coincidence is avoided and conjunctions are brought about between various patterns of movements of the respective elements that map the Ongee cosmology. For the Ongees it is not space and time, or even space-time, it remains only space and movement. The life of the Ongee hunter and gatherer is ordered on the principle of the movement of smell and wind, through which possibilities and probabilities of life are talked about and dealt with. In this 'total' (cf Mauss, 1954) olifactorial and anemological life, and Ongees create situations of inducing and releasing smell and also of conserving and restricting the release of smell. In this lies the continuity of Ongee hunter and gatherer without being hunted and gathered; a balance of life and death between humans and spirits.

In our 'economy of sentiments', emotions and experience, colours remains the prisoner of form, but not sound and smell. Both the sound and the smell of an object always escapes—it is an active principle. Sound and smell are for us like space, evoking a notion of freedom. Smell is distinguished by formlessness and sound is put in a series to make a form, distinct from noise. We may have a problem with the indefinability and lack of articulation of the formlessness of sound and smell (Gell, 1977). Thus bottles of wine are to be opened to sniff, advertisements of colognes (capable of seduction) are to be scratched, and we have to be concerned about the situation that stinks, aroma that fills, whiffs that flow, as well as techniques to make the olfactory functions easy through decongesting medication. For the Ongees smells are complete within the source and so highly concentrated that it becomes substance like bone. Apart from this, for the Ongees the smell is completed in terms of the capacity to 'fill up' and 'empty out' the context in which the source of smell exists. In other words, all that moves and is capable of moving within the cosmology, is subject to smells and winds, both of which move together. The only significant difference, as the Ongees themselves say, is that 'Smell is contained in everybody like tubers are contained in the ground or in the basket, but the winds can never be held, the wind moves anywhere and everywhere'. The Ongees do not have meanings for smell by distinguishing other smells (as is the case with signs), but they do talk of winds being distinguished, because with the wind's movement is the movement of smells and spirits. Thus the olfactory and non-olfactory context for the Ongees is non-existent, since in all contexts the winds and spirits are present and capable of moving. In fact, the spirits and winds for the Ongees acquire the position of sound, the quality of being ever-present becoming almost intense. Just as not having any sound is also a form of music, not having a smell is a way of preventing the spirits and winds from moving close to humans.

In the Ongee world, the experienced cosmology and the mapped movements are thus a discourse on smell and winds, much like the way we talk about sound (formless) and colour (bound by form). For the Ongees smell is bound to form in each place, and the wind is formless, subject to space. Place, where the movement is reduced to a point of security, safety, and nurturance, is opposed to space, which is associated with freedom danger, and uncertainty (Tuan, 1977). It is our notion of colour that forms the Ongee notion of smell, and it is our notion of sound that forms the Ongee notion of wind, through which the Ongees talk about change in durations, temporal units, and time *per se*). Consequently, all movements, in terms of when they will take place, are subject to smell and wind conditions. The conditions of the winds positions the body with or without smell, and determines the interactional outcome between the Ongees, the animals, and the spirits.

When I searched for the answer to the question of when the groups and families will move (a characteristic of the hunting and gathering societies (Service, 1966)), I realized that movement was not a mere translocationary act within a given space connecting places through acts of movement. In the very act, translocation was the dynamics of winds and smell, and an outcome of this was the patterns of social actions and forms of social interactions. Within Ongee cosmology, human beings and animals have limited potential for movement, and are restricted to movement along the horizontal axis of space to places like the forest and the sea. In relation to this, birds and spirit communicators are different. They, with animals such as the monitor lizard and the civet cat, are forms of life that can move across land, into water, and also upwards. Within this classification of various elements of Ongee cosmology, based on the elements' capacity to move, all humans and animals are in lower positions when compared to the hierarchical positions occupied by the spirits, winds, and birds. Spirits and winds are elements of particular significance, that can move any and everywhere within Ongee space. Within this system of the hierarchy of movement, a certain amount of flexibility and change of position is attributed to each element and is visible through the emission of smells, which are carried to the spirits who are the prime absorber and receivers of smell. This movement of smell and wind sets up the possibility of equality within the hierarchical system. The equality is evident in that, though the spirits' movement is greater in its range, by absorbing the smell of humans the spirits transform humans into spirits. When the spirits absorb the smell contained in the human body, they cause death. In the same way, spirits become humans through the process of birth, in which the condensed smell form of the spirit is absorbed by women. In the Ongee world-view and in their ontology, spirits and humans, though hierarchically differentiated in terms of their capacity to move, also become equal through the movement of smell.

The movement of smell sets up the basic quality of dynamics and distinction within the hierarchy of elements such as human beings, animals, and spirits within the Ongee world. The relative position of elements within the hierarchy depends on the characteristic opposition of smell absorbers and smell emitters. Each element characterized as living limits its dispersal of smell. Loss of smell transforms living elements into dead elements not capable of emitting smell but only absorbing it. However, all living elements by the loss of smell are capable of becoming non-living/dead elements. This makes it possible for the elements within the hierarchical system to change their relative position, depending upon whether smell is kept, dispersed, or completely lost. This makes it possible for the spirits to become human beings and human beings to become spirits from the Ongees' point of view. A basic concern for this is evident in day-to-day activity as well as marked moments of rituals within the Ongee camps. This interrelationship of elements is further clarified in the Ongee point of view expressed as, 'Spirits hunt and gather in our island, if we are not hunted and gathered by the spirits then the animals get hunted and gathered by us!'

To write about a world like this, I have taken the idea of smell and winds to be the interconnective, underlying theme throughout this ethnographic account. I intend to show that, while the winds affect and connect with smell to create different situations in the life of the Ongee individual and the community, the basic day-to-day praxis of the Ongee hunters and gatherers is not different from the distinctive duration of ritual. The way power, through the release and absorption of smell, sets up a hierarchy as well as an equality between humans and spirits is replicated on ritual occasions, such as the initiation of young men. The Ongees and the spirits coexist throughout the time span, and the Ongees have to deal with the spirits, dangers, probability, possibilities, and safety. They are both hunters and gatherers, and therefore the concerns for movement are common to all contexts. In all situations, the concern to affect smell and the winds forms the basis for determining the outcome of the interaction between the spirits and the Ongees. This forms the guiding principle for the structure of this ethnographic account and analysis.

As the days went by, and no big spectacular ritual was held, I started hearing and observing more carefully. I felt that perhaps my study of ritual was a wrong question to be brought to the Andaman Islands, just as Radcliffe-Brown could not get genealogies or formulate any argument about pre-totemic social forms (Langham, 1981: 244–7). Since there is so little of ceremony and ritual and lack of structural features that constitute the usual anthropological focus of anthropological writing, I decided to understand values and beliefs more from talking than from observing. Once I understood something of their modes of thought, actual practice became visible in a new light. Every 'normal' activity had to be acutely observed and questioned. It was soon apparent that Ongee activity was far from being devoid of ritual, but symbolic significance had to be looked for in the mundane activity of keeping fire, preserving bones, and the use of clay paints on the body. Concern with fire around the body, bones from dead bodies, and bodies covered with clay paints had a regularity and pattern which were important for the hunting and gathering individuals. I realized that there were few actions that were not is some way prescribed along with the notion of repercussions; so that in order to understand Ongee modes of thought, one must focus upon the way they act in their everyday tasks. Only by living in very close contact could one discern what the actions were and how they fitted in with Ongee notions about themselves, cosmology, and spirits. By constantly being involved in work with them and constantly being questioned about ways of doing the most ordinary tasks, the Ongees with whom I lived became aware of what they took for granted, *totekwata* (old time ways, or traditions) and eventually began to take pleasure in pointing out to me how and why they did something that I had not witnessed or should witness. However, it should not be assumed that native exegesis were ebulliently and effervescently forthcoming. I tried to keep an open mind and be aware of the danger of imposing my theoretical ideas on practice and utterances. There were many evenings in the forest camps when I read translations of myths from the writings of E.H. Man and Radcliffe-Brown to Ongees sitting around a fire. After all, a field ethnographer has to remember that what one hears may not always be the same as what is said and has been said. To a certain extant this account of the Ongee represents an amalgam of what I was told by the Ongees. Remarkable conformity in the different people's response and explanation stems from the fact that Ongees are a small group and there is a lack of social stratification, making knowledge a constantly exchanged thing among all the members of the group. Throughout this book I try to let the Ongee speak for themselves. Whenever possible I refer to their interpretations or justifications. In many instances a more detailed and extensive analysis could have been attempted but the major purpose of the book is to present a general ethnography of Ongee society.

   This ethnographic text has been composed with the intention of conveying how things in a culture come to be known to the people within the given culture and the ethnographer in the given culture. Consequently this is a study of the principles that govern the way the Ongee act, based on their understanding of themselves, their environment, and the supernatural. This ethnographic account is based on my year long fieldwork (1983–4) among the Ongees of Little Andaman island. The ideas and concepts learned from the residents of Dugong Creek have been revised over a period. In writing up, I let the material dictate my approach. Guided by what was made significant to me and how, is the basis of the scheme for the book. Part One is concerned with whole range of relations that are important in relation to the island of Little Andaman, which the Ongee call Gaubolambe. It also includes the relation of earlier ethnography and it's analysis and how it obfuscates or illumines the Andamanese and more specifically Ongee culture. In part two, Gobolagnane's wider social universe in which society and cosmology coexists has been considered through interpretation of Ongee actions using magical substances. The spirits involved in daily life, via the medium of the rules in Ongee culture constitute a theory of causality and movement, specially in relation to smell. The explanations of magical substances smells and movements are internal to the whole of the Ongee cultural system. Part Three, *Gikonetorroka* uses the ideas from first two parts about various relations within the Ongee view of the world in explaining the role of rituals in Ongee culture. Part Four, *Tanageru* is a presentation of the actual initiation ritual in which various Ongee concepts as outlined in the previous parts is used to analyse and explain the ritual of young men's initiation.

Soon I started hearing about the 'sending away of the children and then bringing them down'. On my asking what this was all about, I was generally told, 'You will see when we come to it, the boys will become better husbands and fuller Ongees capable of moving in the sea as well as in the forest along with the spirits!' This was the ritual of initiation of boys known as *tanageru*. However, as is typical of the Ongees, they could never say definitely when the ritual would start. I had a research permit to be on the island for a limited time that was coming to an end. My supply of rations and raw stock of films, also my health after several bouts of malaria, were all on the decline. The prevailing rough sea and storms restricted further delivery of supplies after the month of June. When the Ongees got round to performing the *tanageru* ritual, I realized that all I

had observed on the days preceding the ritual was important.
It was like learning a grammar to understand the fine literary
statement in the ritual.

   This study therefore sets out to deal with the grammar of
movements within the Ongee space and then shows how the
grammatical ideas are used in ritual. The sequence of the
chapters is therefore structured and flavoured in the way the
Ongee culture was approximately understood by me. I started
with questions like why is this ritual organization of these
'symbols' used in this society for these ostensible purposes,
which are perhaps 'our' or a 'reader's' questions. They were
not the questions understood by the residents of Little Andaman. I intend to show why certain ideas and things in relation
to actions pertaining to movement, power, smell, winds, and
hunting get organized as 'Symbols' used in Ongee society for
the ostensive purpose of initiating boys, so that while living
'within the forest' there is also a concern with 'above the forest'.

Dept. of Anthropology,   VISHVAJIT PANDYA
Victoria University,
Wellington,
New Zealand

Acknowledgements

An anthropologist's fieldwork is marked by memories of journey. Thinking about those memories, one realizes that the journey started long before reaching the actual location of fieldwork. One also thinks about all those people and events who made contributions to both the journey and the memory of that journey. My fieldwork among the Ongees is a long journey for which there is no one temporal point marking either its beginning or its end. I would like to acknowledge and thank the many people who have helped, contributed, supported, and criticized my memories and made them memorable.

I wish to thank the Government of India, Andaman Nicobar Administration, the Anthropological Survey of India office at Port Blair, and Andaman Adim Jan Jati Vikas Samiti at Port Blair for making it possible to undertake field research in the Andaman Islands. The research and my study was made possible through grants from the American Institute for Indian Studies, Department of Anthropology and Committee on Southern Asian Studies at the University of Chicago.

I am grateful to Ralph and Marta Nicholas who gave me the intellectual freedom, support, and courage to conceptualize, and think about Andamanese culture as a study project. Many faculty members at the University of Chicago, specially Jean Comaroff, Nancy Munn, Valerio Valeri, Paul Friedrich, McKim Marriott, and Marshall Sahlins have helped me in innumerable ways. Milton Singer and the late Fred Eggan showed me the various ways to appreciate the project of 'salvage ethnography'. Comprehension of the Andamanese culture, specially its linguistic richness and uniqueness, would have been inaccessible to me without the help of teachers and friends like Norman Zide, Gerald Diffloth. I am especially

grateful to Raymond D. Fogelson who has always shown me
the right spirit of anthropology as teacher, consultant, and
warm friend. Colleagues like Lee M. Kochems, Michael Sul-
livan, William Sax, Aditi Nath Sarkar, John Kelly, and John
Levitt, with their comments and suggestions, have contributed to and enhanced my understanding of the ethnographic
material and 'anthropological experience'. In innumerable
ways my parents made it possible for me to experience the
study of anthropology, and specially my wife, Roxanna, without whose encouragement and help this book would not have
materialized. A special word of thanks are due to my editor at
Oxford University Press for his patience and care.

Above all my greatest debt in this research is, of course
owed to the Ongees of Little Andaman island, particularly
Totanage who has been my teacher and often like a big brother;
Tai, Berogegi and Teemai for teaching me about the forest;
Koyra and Muroi for introducing me to the things above the
forest.

V. P.

Contents

Illustrations

Plates (between pp. 104 and 105)

1. Coastal area near Dugong Creek.
2. Maternal aunt carrying her nephew after his face has been painted with clay by his mother.
3. Typical *korale* set within the forest.
4. Returning from a successful pig hunt.
5. Initiates on the way to the forest to hunt pigs.
6. Novices returning after hunting pigs.
7. Initiate sitting on the carcases of pigs during the *tanageru* ritual.
8. Shooting arrows into the pig's skull in the course of the *tanageru* ritual.

A Note on the Orthography

The orthography I have used for the Ongee language is based (with some alterations) on a scheme introduced by Dasgupta and Sharma in *A Hand Book of Onge Language* (1982); for further discussion see Zide and Pandya (1989). The following list gives the nearest English equivalent to the letters used.

a   pronounced like 'a' as in fl*a*sh.

o   pronounced like 'o' as in h*o*t.

i   pronounced like 'i' as in f*i*ght.

u   pronounced like 'u' as in p*u*t.

e   pronounced like 'e' as in p*e*n.

k   pronounced like 'k' as in *k*ey.

g   pronounced like 'g' as in *g*ift.

n   pronounced like 'n' as in ri*n*g.

c   pronounced like 'c' as in *c*hurch.

j   pronounced like 'j' as in *j*udge.

t   pronounced like 't' as in *t*ea.

d   pronounced like 'd' as in *d*ay.

b   pronounced like 'b' as in *b*ook.

m   pronounced like 'm' as in *M*ay.

y   pronounced like 'y' as in *y*es.

r   pronounced like 'r' as in *r*un.

?   Glottal stop. To pronounce this sound the air-stream is stopped at the glottis and released abruptly.

PART ONE

Gaubolambe

CHAPTER 1

Ongees, One of the Remaining Groups of Andaman Islanders

INTRODUCTION

This is an account of the hunting and food gathering tribe of Ongees living on the island of Little Andaman, located in the Bay of Bengal. The 238 square mile island of Little Andaman has a sandy coast-line that blends with thick forest inland. About fifty years ago the Ongee camps were scattered all over the island. Today the Ongees are confined to the two main areas of Dugong Creek and South Bay. However, all the Ongees work their way through the fairly dense undergrowth of the forest and continue to visit the old camp-sites. The entire island is criss-crossed by small streams. In some areas the streams are infested with crocodiles and have to be crossed either by wading through or by walking across on a tree trunk. The south-western and north-western monsoons create a rainy season which lasts approximately nine to 10 months each year; annual precipitation is 2,750 to 4,550 mm. The only dry season on the islands begins in February and ends in March. Except when the weather is dry over an extended period of time, the forest bed is damp and exudes a smell of decaying vegetation. After a heavy downpour the streams are flooded, the paths become slippery, and leeches and ticks abound. In the forest there is a mysterious stillness and silence camouflages the presence of wild pigs, monitor-lizards, civet cats, and snakes, as well as various types of insects and birds. In places where the streams flow into the sea, there are large areas covered with mangrove forests, where plants and animals such as crabs, fish, stingray, mudskippers, and water snakes have adapted to the constant changes of water level caused by the tides. In this study, I propose to show that the life and technology of these hunters and gatherers, constituting their 'praxis', is based on their 'epestime'. The Ongee tribesmen do not simply pick up their 'limited household belongings and move from one place to another in search of food'. According to their cosmology, the space in which hunting and gathering are done is also visited by the spirits, who hunt and gather just as the Ongees do. Moreover, the spirits hunt not only animals but also the human Ongees. This conception has important implications for the role of nature in Ongee culture, and for culturally governed actions within nature. For the Ongees, moving from place to place in search of food places human beings in a vulnerable position *vis-à-vis* the spirits, since the spirits can hunt and gather the Ongee. Such encounters between humans and spirits lead to events that are considered either good or bad, since contact with the spirits, through the dispersal of smell carried by the wind, can lead not only to the death of an individual Ongee, but can also bring him close to the spirits. When an individual comes close to the spirits and is still capable of returning to human society, he acquires power that is intrinsic to the spirits and also becomes an accomplished cartographer and navigator within Ongee space.

The Ongees call their home island Gaubolambe. Even in past times[1] of apparently regular inter-island visitation, Gaubolambe was exclusively Ongee territory. The first known external contact with the Ongees was made in 1885 by a group of British surveyors under the leadership of M.V. Portman. The early encounters between the Ongees and the British were by no means cordial, and the Ongees gained an enduring reputation for overt brutality.[2] This hostility was apparently caused by fear of the outsiders' firearms. Nevertheless, the Ongees did

[1] In 1908, the term Andamanese referred to thirteen distinct tribal groups each distinguished by a different dialect and geographical location. Today only three tribes apart from the Ongees remain and are collectively referred to as 'Andamanese'. The other three extant tribes are: Ongees of Little Andaman Island, the Sentinelese of North Sentinel Island, the Jarwas of the Middle Andamans, and the Great Andamanese of Strait Island.

[2] See M.V. Portman, *History of Our Relations with the Andamanese*, vol. 2 (Calcutta, 1899), pp. 768-808, for historical details.

receive some outsiders, such as the Burmese traders who collected Nautilus shells, birds nests, and ambergris. In exchange, the Ongees received bits of scrap-iron to make tools, as well as opium, tea, clothes, and tobacco. The Ongees also began canoe expeditions to trade as far away as Port Blair. Relations with the outside world, together with socially transmitted disease, contributed to the steady decline of the Ongee population. In 1901, there were as many as 672 Ongees according to the census report of Lt. Col. Sir R.C. Temple (1903).[3] This relationship between the 'native' and the 'outsider' has characterized the history and the history of depicting the Andamanese.

The Andamanese are believed to share a cultural affinity with some of the Orang Aslis of insular South-east Asia. It has been argued that the Andamanese arrived from the Malay and the Burmese coasts by land in late tertiary times or, later, by sea. There is also speculation that the Andamanese came from Sumatra via the Nicobar Islands. However, the precise origins of the Andamanese remain scholarly speculations which have not been thoroughly investigated and researched. The early recorded history of the islands began in earnest with the British in 1788. Rapid changes in trade winds in the area, monsoons, and corral reefs surrounding the islands caused many shipwrecks; those few who survived a shipwreck were killed by the Andamanese. In an effort to establish a safe harbour for their ships, the British made many unsuccessful attempts to pacify the islanders. In 1859, the British established a penal settlement on Middle Andamans; the location was chosen because it was

---

[3] In 1800, the total tribal population on the Andaman Islands, an archipelago consisting of 348 islands, was estimated at approximately 3,575. In 1901, the estimated figure dropped to 1,895. and in 1983, the total tribal population was 269. Of the 1983 estimate only the count of 9 Great Andamanese and 98 Ongees was accurate. The Jarwas and the Sentinelese are isolated by topography and by each tribe's hostility towards outsiders. Since 1789, the population of non-tribal peoples on the islands has steadily increased. The total number of outsiders present on the islands in 1983 was 157,552 compared to 269 tribals. The intrusion of outsiders and diseases introduced by them such as measles, ophthalmia, and venereal disease, have directly contributed to the overall decline in tribal population and its disproportionate male/female ratio. The islands' expanding timber industry and the settlement of increasing number of non-tribals, primarily from mainland India, also have reduced the total area available for use by the tribals.

fortified by its isolation and by Andamanese hostility. Over a period of time the Great Andamanese, who occupied the forests surrounding Port Blair, were pacified and even co-operated with British authorities in tracking down escaped convicts. The British colonial administration established 'Andaman' homes for the tribals in an effort to foster a cordial relationship via exposure to European civilization. By 1875, Andamanese culture came under scientific scrutiny as it was realized that this was a group of people who were dangerously close to extinction. From 1879, under the direction of British scholarship, Andamanese culture was documented, cata-logued, exhibited, and written about, especially with regard to linguistics and physical anthropology. Since Indian indepen-dence in 1947, many different plans for the economic, social welfare, and development of the islands and the tribal popula-tion have been implemented. Today the islands form a part of the Union Territory of India and the remaining four tribal groups are under the government-controlled institution called Andaman Adim Jan Jati Vikas Samiti. Government planners, administrators, and social workers face a dilemma in determin-ing what kinds of changes in the traditional world-view of the remaining tribal groups, especially the Ongees, should be ef-fected. The Jarwas and the Sentinelese have largely remained outside the framework of structured and prolonged welfare activities. The Great Andamanese who, of the four groups have had the longest period of contact with outsider, are the most dependent on outsiders and their goods; they also are the smallest group with practically no memory of their own lan-guage and traditions.

   In September 1983, when I started my field-work among the Ongees of Little Andaman, there were 96 Ongees dis-tributed in the areas of Dugong Creek and South Bay. In the course of my field-work I witnessed two childbirths, bringing the population upto a total of 98.

Ongee Myths and History

Today, outside contact continues to be severely limited and the Ongees live much as their ancestors did: the spirits are feared and most of life is conditioned by the cycles of winds and seasons. Tambolaie, one of the 'Rajas' on Little Andaman,[4] related the following narrative[5] about his people, their past and their contact with the outside world:

Myth No.1

Many moons ago after the winds of many directions went by, from Jackson Creek people of Nappikuteg's band made a big canoe. Many people could sit in that canoe. The seasons were good then—many men and women with children went to the island of Tetale and even much beyond that. Women and children would collect turtle eggs—the spirits got angry because women and children collected so many eggs. [Unit. A.]

So the spirits decided that only stones would be left at the island of Tetale. We knew from the spirits that the Ongees were not to have the turtle and turtle eggs in the same season. [Unit. B.]

Since the Ongees have a big and good canoe they all decided that they would go beyond the island of Tetale—they all went to the Aberdeen and returned with tea and tobacco. It had very tasty smell and smoke. We all give up our forest's leaves for smoking and drinking. Women told the men to go to Aberdeen again and again to bring leaves of various sorts. [Unit. C.]

Then one day some young women went along with the men. The spirits got very angry at this. Our forests were full of leaves but the Ongees did not take them. The spirits got more and more angry. With lots of winds from all directions all the spirits came down and they had a war (*kugebe*) with the Ongees. After the war ended all the Ongees were made into stones (*kuge*). The winds were so harsh and the spirits were so angry and strong that the canoes on the way from Aberdeen were all overturned. The people and the canoes all became *kuge* due to that *kugebe*. [Unit. D.]

No one came back from that trip. Young boys left at home grew up without any friends. So the young boys decided to marry old women, and then the boys had to marry fathers' sisters and girls from the father's sisters' home. [Unit. E.]

---

[4] The term 'Raja' was introduced to the Andamanese, including the Ongees, during colonial rule. From the Ongee point of view a 'Raja' is a responsible and respected spokesman for the outsiders.

[5] Ongees refer to narratives like this by term *jujey*. Following Levi-Strauss's ideas about logic in mythic thought (1963: 206-31), *jujey* could be regarded as 'mythology'. In Ongee *jujey* there is also the meshing of 'perceived contexts' and 'received categories' or 'practical reference' and 'cultural sense' (Sahlins, 1985: 144), making the *jujey* a discourse on history.

Many of the white men followed the rocks formed in the sea due to *kugebe* and they all came in the big boats to Bommilla Creek. At Bommilla the white men found only Ongee men. No one had women. White men wanted to give leaves to Ongee men and women. On not finding Ongee women they did not give any leaves. White men wanted to take Ongee men and women away to Aberdeen and they had all come because the spirits helped them to come to the Bommilla. It was all a bad coincidence (*malabuka*). Once again there was a big war. Many were turned into *kuge*. [Unit. F.]

After this, when the young boys grow up they have to marry within their fathers' sisters' home. This is the way the Ongees go from one relative to the other and avoid encounter and war. [Unit. G.]

When all the rocks and stones go away from the sea, like the ones we see now between us and Aberdeen, then nobody will go or come—no smell will move. We have our leaves to drink and smoke. All the pain will go away. No longer will the spirits get angry—our fathers and mothers will all remain with us. We will keep no lower jawbones with us because the spirits will not come to us. Our own teeth will not fall down. No one will die and there will be no spirits. There will be no marriage and giving of young boys to the old and the spirits. [Unit. H.]

We give the boys to the turtle hunters group and turtle hunters' boys are given to pig hunters. When all the rocks are washed away, many young boys and girls will be living. We will have no need to run and hide from the spirits. Turtle hunters (*eahambelakwe*) and pig hunters (*eahansakwe*) will have boys and girls for themselves. There would be no death and the need to marry outside. Everyone will have a husband and no one will have to go to his wife's band for marriage. [Unit. I.]

The Ongees narrate the following myth to explain what happened before they had contact with the angry spirits and white men. This narrative not only describes the situation before contact with the outside world, but also tells about the origin of the Ongees on the island of Little Andaman.

Myth No. 2

Dare the season of the south-east winds associated with the spirit called Tenneyabogalange, brought two huge bamboo nodes. Tenneyabogalange found them in the sea of Lauterka and she took them to her own home. The bamboo nodes came from two different places. One came from the north-east and the other came from the south-west. [Unit. A.]

It was hot and Dare was restless, causing much of the winds to come on our Ongee land. The bamboo nodes cracked open due to the heat and loss of moisture. Inside the bamboo nodes were two spirits—Mayakangne? and Kwalakangne?. As the two came out there was rain all round—water everywhere—and as the rain stopped Mayakangne? and Kwalakangne? had grown up into adults. They grew so rapidly that the home of Dare became small for them. At this Mayakangne? and Kwalakangne? were instructed by Dare to go down to the land of the Ongees one by one, and in turn eat whatever was to be found and was liked by them. Mayakangne? and Kwalakangne? started living on this island, and they kept growing bigger and bigger. [Unit. B.]

One day they both succeeded in pulling a huge turtle, that was initially hunted by Kwalakangne?, to the land. Once the turtle was pulled to the shore, it was dragged all the way to a nearby anthill and left there to die. The turtle lost all its heat and started acquiring moisture. At this point Mayakangne? and Kwalakangne? decided to slit open the turtle. Inside the turtle shell was a woman and her sister. [Unit. C.]

Kwalakangne? and Mayakangne? married the two women who had come out of the turtle. Out of these two marriages each couple had a daughter—Eneyagegi ('women's grass apron') and Eneyabegi ('milky breast of the women'). Eneyagegi and Eneyabegi were not to remain on the island of Gaubolambe all alone since their parents had outgrown their size. They were too big to live on the island. Mayakangne? and Kwalakangne?, along with their respective mates, left for the home of Dare. They promised that whenever the children felt lonely and hungry they would come down from Dare's home with the rain and wind. [Unit.D.]

Eneyagegi and Eneyabegi were very lonely and they sat waiting for Mayakangne? and Kwalakangne?. They built a shelter on the coast and started keeping food on the thatch for their grandmother Dare and parents Mayakangne? and Kwalakangne? along with the winds and rain. Winds and rain were their brothers and sisters. They wanted their parents and grandmother to know that they had food to give and share but they were very lonely and wanted them to visit the island. [Unit. E.]

While waiting, Eneyagegi and Eneyabegi decided to go for a walk in two different directions. Eneyagegi went towards the thick forest and Eneyabegi up along the creek but then turned and walked towards the seashore. Upon coming back to the big shelter they had made, they brought along with them branches of *tukwengalako* [a type of tree, *Dipterocarpus incanus*]. They both sat at the coast with the branches. It was low tide, many crabs were to be seen all around. On a nearby tree a civet cat (*kekele*)and a monitor lizard (*ayuge*) were also sitting. From the sea many turtles came out. All the turtles made holes in the sand and laid eggs on it. [Unit. F.]

Next high tide the turtles came to fetch their young. They were all very unhappy on not finding what they had come for. Seeing the plight of the turtle, the crabs, who were good friends of the turtles, informed them of what had actually happened. After the eggs were laid by the turtles, the *ayuge* came sniffing around, he found the eggs and ate them all. At the next full-moon, the turtles were back to lay more eggs and at the suggestion of the crabs they circled around the spot where they had laid eggs. Lizards who came smelling the tracks of the turtles could not get to the eggs, since they kept smelling the turtle tracks and kept going round and round the holes where the eggs were. [Unit. G.]

At the next high tide the turtles came to see their young and were happy to see that they were ready to come along to the sea. From then onwards the turtles and the crabs became very good friends. Eneyagegi and Eneyabegi saw all this and they decided to put the *tukwengalako* branch in a hole at the place from where they had brought it. One pair of *tukwengalako* branches was placed on the forest side and one pair was placed on the coastal side. After this Eneyagegi and Eneyabegi moved round in a circle around the place where the branches were placed. [Unit. H.]

So the turtles at the suggestion of the crabs had done a dance (*onolabe*) and then the same *onolabe* was done by Eneyagegi and Eneyabegi. Eneyagegi and Eneyabegi were very happy, they sang songs and buried yams (*gegi*). Female *gegi* root was placed along side the forest. After this Eneyagegi and Eneyabegi went to sleep. On waking up they found that at the place where they had planted the two pairs of *tukwengalako* branches stood the Ongees. At the sea-side was a man and a woman and on the forest side too there was a man and a woman. They were the first turtle hunting Ongee (*eahambelakwe*) and pig hunting Ongee (*eahansakwe*). [Unit. I.]

Dare, Kwalakangne?, and Mayakangne? all came down. There was a flood and storm. A big storm and lots of rain. They were angry, since there were now men who were to be hunters of pigs and turtles. So Dare told Eneyagegi and Eneyabegi that they should make it clear to the Ongees that they would have to live along with the winds coming and going along with Dare, Mayakangne?, Kwalakangne?, and all those who die because they have to come down to feed. The Ongee women were to see to it that the men do no bad work and make sure that all the dead and those who move along with the winds get things to eat. After this everybody left the island of Gaubolambe—but the Ongee alone were left. [Unit. J.]

The *gayebarrota* (forest) and Ongee labour in it provided the Ongees with virtually all their necessities. Hunting and gathering, predicated on a seasonal translocationary pattern, characterizes Andamanese culture. The Jarwas and Sentinelese are still completely dependent on hunting and gathering activities.

Among the Ongees, however, plantation cultivation of coconuts was introduced in 1958. Although the Ongees gather the
coconuts they do not want to be involved with nor do they
participate in any form of agricultural activity. The Ongees are
paid for gathering coconuts by the welfare agency with food
rations and industrial products from mainland India. Consequently, their consumption of forest products alone is being
increasingly substituted by imported products. Among the
Great Andamanese hunting is only an occasional activity. They
are paid a monthly allowance by the government and also
receive wages for taking care of the citrus fruit plantations.
Fishing in the sea is usually accomplished with bows and
arrows while standing in knee high water, especially during
low tide, and is a year-round activity. Occasionally lines and
hooks are used to fish in the sea. Hand-held nets are used to
fish and for gathering crabs and shellfish from the island's
inland creeks. Fish is an important part of Andamanese culture;
in the different dialects the term for 'food' is the same as that
for 'fish'. Traditionally the northern groups caught sea turtles
in large nets, but this is not the case with the southern groups
of Andamanese. Ongees paddle out to sea in their dugout
outrigger canoes to hunt sea turtles and dugongs with harpoons. During the wet season the Andamanese hunt pigs in the
forest with bows and detachable arrowheads. Dogs, introduced to the island in 1850 and the only domesticated animal
among the Andamanese, are sometimes used to track down the
pigs. Throughout the year there is a significant dependence on
gathering different items such as turtle eggs, honey, yams,
larvae, jackfruit, and wild citrus fruits and berries. Apart from
hunting, the Ongees forage for numerous species of fruits and
edible tubers, and gather nuts, seeds, and honey. Traditionally
the Andamanese were dependent on the forest and the sea for
all resources and raw materials. Raw material, such as plastic
and nylon cords, have now been incorporated into Andamanese material culture: plastic containers are used for storage,
nylon cords are used as string to make nets. These items are
most usually discarded by passing ships and fishing boats and
then are washed up onto the islands. The Indian government
distributes as 'gifts' to the Ongees, Jarwas, and Sentinelese

metal pots and pans[6] and as a consequence metal cookware has nearly replaced the traditional hand-moulded clay cooking pots that were sun dried and partially fire baked. The Ongees continue to make clay pots but primarily use them for ceremonial occasions. Ongees grind metal scraps, found on the shore or received from the government, on stones and rocks to fashion cutting blades and arrowheads.

Prior to the introduction of metal in 1870 by the British, the Ongees fashioned adze and arrowheads from shells, bones, or hard wood. Although iron is highly valued by the Ongees, they do not use iron nails to join objects. Ongees still join objects by carving or tying rattan rope, cane strips, or strands of nylon cord. Smoking pipes, outrigger canoes, and cylindrical containers for holding honey are some of the many items carved by the Ongees. This makes Ongees somewhat distinct in relation to other Andamanese as far as territorial organization is concerned and its effect on the politics and economy.

Among the Andamanese certain territories were identified as belonging to a specific band. In the Northern and the Middle Andamans it was frequently necessary to pass through another's territory. The trespassers were obliged to behave as guests in another's territory and, in return, the owners of a given territory were obliged to behave as cordial hosts. Thus a feeling of mutual interdependence and a value for hunting and gathering in each other's part of the island created a process of production and consumption which was to be shared. Among the Ongees of Little Andamans, where no other tribal group resides, the island is divided into four major parts and identified with two pairs of mythical birds each of which is associated with land or water. The four divisions of land represent the four Ongee clans. Each section of the island is further subdivided into sections of land associated with a lineage. These land divisions, known as *megeyabarrota*, are identified with a person's matrilineage and, depending on whether the territory is in the forest or on the coast, with either the turtle

---

[6] T.N. Pandit in his book *The Sentinelese* (1990) presents the details of the continued gift-dropping missions. His account makes it apparent that the missions have failed to accomplish anything close to Sentinelese culture's logic but have fostered a dependence of the Sentinelese on outsiders and the crystallization of a peculiar attitude of the outside world towards the native.

hunters (*eahambelakwe*) or the pig hunters (*eahansakwe*). Ongees prefer to hunt and gather in their own *megeyabarrota* but there are no restrictions against hunting in someone else's *megeyabarrota*. If one does hunt in another's *megeyabarrota*, he is obliged to offer and share first with the owners any resource taken. A person's identity with a *megeyabarrota* plays a crucial role in Ongee rituals and ceremonies, e.g. consummation of a marriage must occur in the wife's *megeyabarrota*, a dead person's bones must be kept in the *berale* (circular hut) of a descendent's *megeyabarrota*.

Traditionally, trade within a group was conducted between the bands identified as pig hunters (forest dwellers) and turtle hunters (coastal dwellers). The pig hunters band traded clay paint, clay for making pots, honey, wood for bows and arrows, trunks of small trees for canoes, and betel-nuts in exchange for metal gathered from the shore, shells for ornaments, ropes and strings made from plant fibres and nylon, and edible lime gathered by the turtle hunters. The bands would take turns serving as host for these organized events of exchange. Historically the Andamanese gathered honey, shells, and ambergris to trade with outsiders in return for clothes, metal implements, or even cosmetics. Under the colonial administration trade with outsiders was the point of entry for opium and liquor into the northern Andamanese community.

According to the Ongees, the days before coconut plantation and the help of the welfare agencies, they and their ancestors would travel by canoe northward to Port Blair to exchange with other Andamanese the sugar and tobacco received from the British Administration. Ever since 1952, when the Ongees were introduced to the systematic planting of coconut trees by the Italian anthropologist Lidio Cipriani, 'coconut work' has become another focus of collective labour.[7] The Government of India, through the Andaman Adim Jan Jat Vikas Samiti

[7] In Ongee language doing some sort of work or activity is expressed by the term *ronka*. However doing coconut work, as it was introduced by outsiders and comes with it's own flavour of developing plantations in the tropics and the power relations between managers and workers, is referred to as *totaley* which literally in Ongee language means an unavoidable situation of clean up! It is interesting to note how the Ongee, who have a different idea of work, fit the structure and required work of coconut plantations as *totaley* and not *ronka*.

(AAJVS), with the good intention of protecting the tribe, has set up a permanent residential area for the Ongees on the island of Little Andaman. At these settlement sites, the AAJVS maintains a supply of food and certain commodities such as cloth, knives, nails, and tobacco, all of which are thought to be appropriate by the mainland urban government officials. The AAJVS's effort has been to confine the Ongees to one place. Each Ongee individual covers approximately 2.47 square miles in terms of hunting and foraging activities. This is seen as a 'major obstacle' to the development of the island's forest resources for creating and supporting the matchstick industry. From the local administration's point of view, this is a major problem. The Ongees collect whatever they can and obtain whatever they want from the AAJVS stores, but they have continued to carry on with their traditional way of life oblivious to the so-called attempts made to modernize them. As a politician, sharing a concern with the local administrator, commented:

> See these people are *junglees*! We have to bring them into the mainstream of the nation's development and modernization—our party is concerned—and leaders, like Nehrujee, Indirajee have told us to protect them—you see, to modernize the Ongee—to give them all that we can we have to make them stay at one place—only animals of the wild roam from place to place—the Ongees have to understand our efforts!
>
> [Translated from Hindustani conversation
> at Tehsildar's office, Hut Bay, Oct.1984.]

A certain amount of resentment towards the external presence and interference is evident in the Ongee attitude towards the AAJVS, and this has a historical basis going back to the days of the Raj when the Ongees expressed hostility towards foreigners. Challenging the external agencies and their efforts to make the Ongees settle down, the Ongees continue to move from place to place, cutting through the undergrowth that covers their forest paths, and to the coast and among the coral reefs in their dugout canoes.

Relations and Social Organization

The present small size of the population and the limited

information available on the Northern and Middle Andamans makes it difficult to create a comprehensive picture of Andamanese kinship. Earlier ethnographic accounts present the basic tribal divisions as 'sept', but Radcliffe-Brown's observations lead us to believe that groups came together to ensure friendly relations. On the basis of Ongee ethnographic material and classical descriptions of the Andamanese, it is beyond a doubt that the Andamanese have bilateral descent groups. The nuclear family is the major group around which all activities revolve. The nuclear family includes a married couple's own children as well as any adopted children. By rule, descent among the Andamanese is bilateral. The kinship system is cognatic and terminology, on the whole, specifies classificatory relations. Prefixes are affixed to classificatory terms of reference which also emphasize senior and junior age differences. Marriage is arranged by the elders within the prescribed group, that is, between turtle hunters and pig hunters. A man's patrilineal relatives take gifts and demand a daughter from a man's matrilineal group. Among the Ongees, population decline often makes it impossible for a young man to marry his classificatory cross-cousin and, consequently, he sometimes must marry a much older woman who is his mother's classificatory cross-cousin. Monogamy is a strict rule. An older man or woman who has lost a spouse receives priority for marriage. Levirate marriage is acceptable. Marriage is a highly valued status. Both Man (1885: 137) and Radcliffe-Brown (1922: 29) imply that residence is ambilocal, but some of Radcliffe-Brown's (1922: 78-80) remarks indicate a tendency towards virilocal residence. Among the Ongees a newly married couple stays with the wife's matrilineal relatives at least until a child is born. After a child is born the couple may move to live with the husband's siblings and their families. Men and women inherit rights and obligations primarily from their matrilineal lineage. Tools and canoes may be inherited from the father's side. Divorce is rare and is considered immoral after the birth of a child. Customarily children are given in adoption. The responsibility of early socialization of the child rests with the child's matrilineal relatives. Once a young boy is ready for initiation his training and education become the responsibility of his father and his paternal relatives. After a girl's first

menstruation she is even more closely aligned with her matrilineal relatives. Children of both sexes are taught about the
forest as they accompany their elders on different hunting and
gathering activities. Through play and the making of toy
canoes, bows and arrows, shelters, and small nets, children are
introduced to basic requisite skills.

Traditionally speakers of a dialect resided as an independent and autonomous group in a specific part of the islands.
Each local group was further divided, especially in the Northern and the Middle Andamans, into sub-groups of 20 to 50
people who, depending on the season, lived either on the coast
or in the forest. Marriage alliances and adoptions between
coastal and forest dwellers controlled conflict, which was sup-
plemented by the dictates of the elders. Occasionally neighbouring groups would have a conflict of interests; however,
hostility never escalated beyond the level of avoidance. When
problems between groups arose, women, through informal
channels of negotiation, were instrumental in the resolution of
tension. Resolution was usually marked by a feast in which the
groups in conflict would participate. Between neighbouring
groups with different identities marked by different spoken
dialects, the peace-making ceremony consisted of a sequence
of shared feasts held over a period of time. The colonial administrators of the islands acknowledged the position of influence held by some of the elders and thus titles, such as Raja,
were introduced and functionary chiefs created. The position
of Raja was always held by an elder who could speak the
administration's language, Hindustani. The Andamanese
value system is the basic means for maintaining social control.
Direct confrontation is avoided and 'going away', that is, leaving the source and scene of conflict for a short time, is encouraged. Usually resentment is expressed by breaking or
destroying some piece of property at the camp-site and then
leaving for the forest to stay for few days. When the offended
person is away, other camp-mates fix the destroyed property
and await his return, which is without recrimination. Some of
the features described in classical ethnography continue to be
evident within the Ongee social organization (*see* Basu, B.,
1990). Given the present size of the total population of Ongees,
much of the 'traditional' rules of organizing social relations are

not very distinctly visible, or sound. For instance, many of the eligible bachelors cannot marry because eligible and available spouses are not from the appropriate and prescribed group with whom marriage could be negotiated. However, the Ongee settlement pattern still follows the old traditions. On an average, about five to eight families get together and set up their shelters in a circular pattern. All Ongee families construct their shelters in a circular pattern. The only exception is the house of *obonaley* that is generally outside the circular arrangement. The ground enclosed by the shelters is called *wabe*. In the Ongee language *wabe* is the root verb for crying and heat generation. The two meanings of *wabe* have special significance to the central camp-ground. It is in the central camp-ground that the fire is kept, and there that all gather together and communicate various feelings and plan activities (cf. Radcliffe-Brown, 1964: 34 for layout of the Andamanese camp-ground). All the men who have never been married or whose spouse is dead or from whom they are separated together form a residential unit. They together build shelters separate from the standard circular pattern in which all the families have individual shelters (*korale*) and sleeping platforms (*kame*). In Ongee language[8] the term *korale* means not only a 'home' but literally stands for the reference term for family. *Korale* is constituted by the term *ko* that is the inquisitive marker and the suffix *ale* that means 'children'. Thus, for the Ongee, when they say *korale* they mean not only a home but the nuclear family. *Kame*, the sleeping platform under each *korale*, is associated with the people who sleep together.

[8] A real linguistic connection of Andamanese with South and South-east Asian language areas has not been systematically established. Andamanese as a language family is composed of two main groups: first, Proto-Little Andamanese which includes Ongee, Jarwa, and Sentinelese; and second, Proto-Great Andamanese. Proto-Great Andamanese is further subdivided into three groups: Bea and Bale of South Andamans; Puchikwar, Kede, Juwoi, Kol, and Jko of Middle Andamans; and Bo, Chari, Jeru, and Kora of the North Andamanese. Early ethnographic accounts suggest that each tribal group on the islands spoke mutually unintelligible languages. However, linguistic records, compiled by the islands' administrators and more recent research, suggest a great degree of overlap in terms used by each group. For further detail *see* Zide and Pandya (1989), 'A Bibliographic Introduction To Andamanese Linguistics' in *Journal of the American Oriental Society*, 109.4, pp. 639-51.

Unless and until the host invites you and makes the appropriate gesture, no visitor is expected to sit on another's *kame*. *Ka* stands for the locative 'where' and *ame* means 'arrival of' or 'birth' or 'the birth of a child'. *Ame* is also the Ongee word for the 'bud' of a plant (*see* Myth No.2 for the idea that Ongees originate from plants). So central is the concept of the *kame* and one's *korale* that when undertaking various translocations the Ongee family always takes along its own *kame* dismantled, or at least the mat which forms its centrepiece.

Unmarried and separated men all have one sleeping platform and sleep together under one shelter. They work together and help others in various situations like cutting logs for canoes, and locating and clearing camp-sites. These men living together under a shared roof are regarded as residents of *obonaley*. *Obonaley* also forms the shelter where all men congregate, especially on full-moon nights. It is *obonaley* where all the boys collect and are instructed in various forms of activities like making canoes, bows, and arrows. *Obonaley* is the only residential shelter in the camp-site which has a fire but never is food cooked under its shelter or by its residents. Residents of *obonaley* have access to food from everybody's family kitchen. Therefore, whatever is hunted by them is all given away within various families. From time to time *obonaley* residents provide food to various families and in return they continue to receive cooked food. Within each camp-site, food, hunting weapons, and foraging implements are shared.

Each evening when the men return to the camp with the kill of their hunt and the women return with what they have gathered, all the food is placed in the middle of the camp at the communal cooking place. All Ongees are expected to just pick up whatever food there is irrespective of who brought it and who is or is not related to the person who either hunted or gathered it. The Ongees explain:

> No one gives food and tools since they are never taken away. They are always there for use. Only spirits take away and do not return.

The Ongee language demonstrates this egalitarian attitude that underlies Ongee notions of giving and taking. The term *ka*, which means 'where', also implies 'to give'. The term *ma*, which means 'not here', also implies a refusal to give something.

However, *ka* and *ma* are never used for requesting or refusing food or tools. The terms *ma* and *ka* can only be used for negotiating marriage, child adoption, and magical substances.

Individuals related through marriage or child adoption, just like the magical substances (*gobolagnane*), belong to a family or a group but can be asked for and have to be returned, if asked for, if their original family believe that *manyube*, ('those who demand a male partner in marriage') and *alankarebe*, ('those who demand a child in adoption'), cannot be refused. However, those individuals can be demanded back by the original families if the marriage or adoption becomes 'unhappy'. The actual term used by Ongees is *nangucumemy*, that literally means 'something that has decayed and is giving a bad odour'. Marriage and adoption by demand are viewed as relations of 'giving and taking'; such relations are always between camp-sites, not within the same. Within the camp-site, Ongees see themselves in a relation of *enekuta*, where things are not demanded and do not have to be demanded back.

When the wife is pregnant, she and her husband decide to hold the *alankare* ceremony. This ceremony is generally held after the pregnant wife is acknowledged to have a 'blockage' in her womb due to the spirit trapped and transformed into a foetus that restricts the flow of blood. Ongees hold this ceremony, especially the husband and wife, by appointing the selected couple who will fulfil the role of foster-parents. In other words, the couples elected by the mother-to-be and father-to-be are seen as the people who are responsible for inviting and taking care of the child. The appointed 'adopter' of the child to be born is called *mutarandee*. Generally, every married man is a *mutarandee* to someone. At the *alankare* ceremony the *mutarandee* goes and hunts food and arranges a special meal for the pregnant woman. During the series of feasts held by the *mutarandee* and his wife, they try to keep track of the number of times the pregnant woman vomits after eating at the feast. If the woman vomits much after eating seafood, then the *mutarandee* declares that the child is a male spirit changing into a male child. If the pregnant woman vomits more after eating pig and tubers, that is, food from the forest, then the forthcoming child is declared to be a female child/spirit. On the basis of the account of vomit the *mutarandee* then arranges future plans, that is, if it is a boy then the *mutarandee* has to become the *muteejeye* ('chief initiator'). If it happens to be a girl, then the *mutarandee*'s wife has the social responsibility for bringing up the daughter as her own and complete her socialization.[9]

The *alankare* ceremony is an occasion where the community is made aware of the people selected as the *mutarandee* and *muteejeye*. At the time of the *alankare* the *mutarandee* also publicly acknowledges his role and ritually talks to the spirit to be born as a child. He promises the spirit/child that he will work hard to make the spirit's stay among the Ongees comfortable.

Every Ongee regards it as a great honour to become a *mutarandee*. In order to be appointed, the individual must be married and the husband must have completed his initiation. It is only the initiated male of the Ongee community who can take the role of *mutarandee*. The *mutarandee* brings the spirit to the community of humans as a child. It is also he who has the responsibility at the time of initiation of sending that child up to the spirits in the form of the spirits. Appropriately enough, the *muteejeye* (initiator) and *mutarandee* (adopter) are responsible for bringing spirits to the *korale* and for sending the child to the *korale* of the spirits at the time of initiation. Generally, within a camp-site all the married women are related through a common ancestor who regard themselves to be *etetingege*, 'sisters'. So strong is this feeling that as soon as a child is born in a camp-site, all the women congregate at the shelter of the new mother and, irrespective of their actual capacity, they all breast-feed the newly-born child. Individuals among the *etetingege* who can breast-feed in actuality feed the child for a prolonged duration.

When a child (particularly male) has attained the age of being able to masticate his own food, he goes to live with his *maikuta*, 'mother's brother'. The mother's brother is responsible for socializing and training the child in various skills and techniques. He and the other relatives on the mother's side form a group called *beyagee*. At the time of marriage the boy moves out of his *beyagee* and goes to his father's sister's group.

[9] *See* Radcliffe-Brown (1964: 77) for his observations on the child-adoption customs; *also* ref. to Ray and Ganguly (1962: 368-9).

The residents of the territory where the father's sister lives with other paternal relatives forms the group called *gaakoulotee*. It is within the *gaakoulotee* that a marriage alliance and suitable marriage partner for the boy is found. A suitable marriage implies rules of *angage*, whereby second and third degree patrilineal cross cousins as well as patrilineal aunts are preferred marriage partners. The Ongee interpretation of the rule of *angage* is based on the principle that there are two kinds of relatives. *Gakodeneye* is the group of relatives among whom the feeding of liquids, such as blood at the time of initiation and milk at the time of birth, is essential. The Ongees explain *angage* as the group of one's father from which ego's father 'comes to' marry ego's mother. It also is the designated group from which ego (male) finds an appropriate marriage partner. This makes possible the collapse of ego's *gaakoulotee* into *guleedange* (relatives by marriage). Thus, *gakodeneye* consists of all the relatives who are connected through a matrilineal line of descent and are related through birth.

The other group of relations is *guleedange*, formed by marriage, where the male has to contribute in terms of work and the products 'of cutting, carving, and binding'. Thus, the rule of *angage* makes it possible for the women to stay at a place identified with either a turtle hunting group or a pig hunting group, which is one out of the four territorially associated clans. Males alone move between the four clan territories for marriage, as well as between their mother's relatives and father's relatives. All the Ongees are associated with one of the four clans. Each clan has a particular sector of the island where it locates its identity through four distinct birds. The fourfold division of the island into territories that form the 'clan identity' is attributed by the Ongees to the places where Eneyagegi and Eneyabegi had planted pairs of *tukwengalako* stems. The four territories had four resident birds known as Amiya, Choulung, Amie, and Gaye?. These birds were appointed by Eneyagegi and Eneyabegi, creators of the Ongees, to inform the spirits whether Ongees were moving in accordance with the arrival and departure of the spirits and winds, so that everyone had something to hunt and gather. Within each fourfold clan division is the twofold division of the *eahambelakwe*, turtle hunters, and the *eahansakwe*, pig hunters. Residents of each territorial unit are divided on the basis of the matrilineally associated part of the land, known as *megeyabarrota*. If the *megeyabarrota* is on the forest side, the identity of the division is that of pig hunters, and if the location of the *megeyabarrota* is on the coastal side, then the identity is that of turtle hunters. All marriages have to take place between the pig hunters and the turtle hunters, and across territorially identified and proscribed clans (*see* Figure 1—Pattern of Alliance and Marriage Exchange).

Every woman has only one identity. She is a member of either a turtle hunting group or a pig hunting group, who has the right to one of the four territorially identified clans. The men, however, have rights and an identity in two out of the four territorially identified clans. After marriage, the man associates with his wife's *megeyabarrota* for the purpose of work, living, and his children. But the same man, even after marriage, identifies with his mother's *megeyabarrota* for training his sister's children, and also because that group provides a wife for his son. Ego movement out of his *beyagee* (matrilineal group)

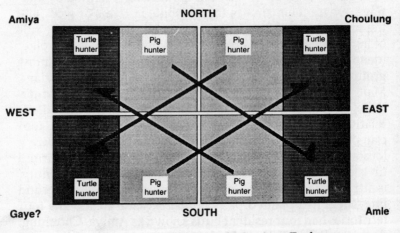

Fig. 1 Pattern of Alliance and Marriage Exchange

towards his *gaakoulotee* (patrilineal group) makes it possible for him to find a wife and form a *guleedange* (group of relations through marriage).

Turtle hunters, *eahambelakwe*, are regarded as people who have an acute sense of sight/vision, referred to as *gebogelah*, 'those who see well'. Pig hunters, *eahansakwe*, are regarded as *gekalegebaro*, meaning 'those who hear well'. Thus, at the time of the marriage alliance, the boy's mother's brother goes to the *berale*, circular shelter, built by the boy's paternal relatives on their *megeyabarrota*, and presents certain gifts, marked by the meat of the animal with which the boy and his mother's brother are associated, and says, 'This is what we can do, we want somebody to demand this boy so that hearing and seeing is completed through *enengelabe*, marriage between the two *megeyabarrota*!' If the proposition is accepted, then, at the wedding ceremony, the turtle hunters and pig hunters bring the two distinct types of meat for a large feast, and the boy starts to live with his wife's matrilineally related group. After the *enenegelabe* the husband and wife set up a new shelter along with the other relatives within the associated clan territory. After the birth of the first child, the husband accepts his adopted children (sisters' boys and/or girls of age four and above), and they come and live in his shelter. Within a given territory, all the turtle hunters host the pig hunters of the same territory throughout the turtle hunting season, and then become guests at the pig hunters' shelters throughout the pig hunting season. The relationship between the turtle hunters and the pig hunters in a given territory is called *gotorokabe gabeyebe eneku tata*, meaning 'together we visit each other and give'. Thus, within the Ongee social organization, no permanent group formation and residence is to be found, especially given the present size of the population and the rule of *angage* which obliges many eligible individuals to remain single. The kinship system being cognatic, the terminology on the whole specifies classificatory relations. The chief social unit remains a nuclear family, several of which usually live together in a settlement, but the composition of residential units changes, reflecting the structural principles of fourfold and twofold divisions underlying the relations within and between the groups of Ongees. Furthermore, there are no leaders in any formal and traditional sense. The exceptions are those who can and do give more in terms of food and labour and are respected along with the elders in the community.

During the colonial period, and now, when the Government of India officials visit the island, the Ongees recognize a person whom the outsiders have appointed as a Raja. The primary function of the Raja has always been that he can talk on behalf of the Ongees to outsiders and outsiders can talk to the Raja in Hindustani. Within the clusters of self-sufficient nuclear families, the only category of person that constitutes influential specialization, and in some sense transcends the social order, is the *torale*, the spirit-communicator. Moving from place to place, *eyolobe* is the underlying principle for the formation of social relations. As noted earlier, relations of marriage and adoption are all founded on individuals and groups moving across territories. Availability of food resources from time to time is also determined by movement of spirits and winds. The arrival and departure of spirits with winds from different directions creates, according to the Ongees, their experienced seasonal cycles. In response to the movement of spirits and winds, the society moves from place to place for hunting and gathering.

Seasonal Changes and Social Responses

The Ongee seasonal cycle (*monatandunamey*) is not only a reflection of temporal divisions, wind conditions, and availability of food resources, but also maps out and provides an itinerary of movement for man and spirit. The Ongees and the *tomya* (spirits) move in relation to each other within the shared space, which makes possible the identical activity of hunting and gathering for man and spirit. The Ongees move from the coastal area to the interior of the forest with the increase of temperature. Thus, from November to February, the pig hunters come to stay with the turtle hunters. This movement is called *gayakabe*, and it takes place in the season of Mayakangne? It is in this period of the year (November–February) that the spirit of Mayakangne?, along with the winds coming from the north-east, comes to the island of Little Andaman. The Ongee say that these winds have an impact so strong that trees fall and

fruits are scattered to the ground in the forest. This phenomenon is called *naregero*. According to the Ongees, in the Mayakangne? season, pigs cannot put on weight because the spirits
visiting the island enter the fallen fruits and the bodies of the
pigs, feeding on them as nourishment. The spirits of Mayakangne? are believed to be actually within the fruits and the
bodies of the pigs. Consequently, pigs become a food item to
be avoided (*erotakabe*) by the Ongees because the *tomya* of
Mayakangne? are consuming them from within. As the *tomya*
of Mayakangne? continue to feed, their hunger and anger
subside, resulting in the decrease of the north-east winds. The
season of Mayakangne? comes to an end because the spirits
have 'sucked' away all the available nutrition from the fruits
and the bodies of the pigs. The end of Mayakangne? is realized
in the decline of rain and moisture, which creates *ikatatabe*,
space made dry.

With the end of the season, the turtles go away from the
coastal areas due to the lack of rain and the resulting dryness.
So, the Ongees leave the coast and return to the forest. Thus, in
Ongee perception, turtles are the food that men can hunt in the
season of Mayakangne?, while the pigs are consumed by the
spirits. The tail end of Mayakangne? is also seen as the season
in which *gegi* (tubers under the ground) start growing. The
season of Torale (middle of March to late April), the peak of the
dry period, follows Mayakangne?. During this period the Ongees collect *tanja*, honey. As the Mayakange? comes to an end
and the season for honey collection sets in, there is a transitional
phase of cyclonic conditions known as *gingetigye*. During this
period, women start dismantling and packing the shelters set
up along the coast. People also busy themselves preparing for
the ritual of *getankare*, which involves carving wooden phalluses and selecting trees for carving out canoes. Once the ritual
of *getankare* is over, the Ongees divide into family units and,
with their pack of dogs, proceed to the forest. This marks the
formal start of the season of Torale.

Once the ritual of *getankare*, involving various operations
of offence, has been performed by the Ongees, they are sure
that for a short duration it will remain a man-made season. Just
as people carve and extract things, so do they make the dry
season of Torale, in which the spirits are absent from the island.

This season of Torale delays the arrival of the spirit Kwala-kangne?, associated with the south-west monsoon. Unlike the previous stay of Mayakangne?, which makes the pig a food item to be avoided, the absence of the spirits from the island makes honey available to the Ongees. Honey is spirit food that the spirits most favour and one they are always consuming. For humans to have honey they have to create a short duration when the spirits are absent from the island.

By the end of the second week of March all the Ongee families, one by one, start moving towards deeper forest. The order in which the families pack up from the coastal areas reflects the status of the elders, who are privileged and leave first. Younger couples leave only after the elder couples have settled in the forest. It is the responsibility of the elders to move into the forest and locate trees with honeycombs and choose the location for the shelters. Residential areas are always selected with two factors in mind: (i) there should be fresh water nearby, and (ii) no dry tree nearby that can catch fire.

The Ongees prefer to locate in the section of the forest where they have preserved the bones of their dead ancestors under a *berale* (shelter). Each family maintains a *berale* in the forest where all the ancestral bones are preserved, and the shelters are maintained throughout the year. The section of the forest where the bones are kept in a *berale* is called a family's *megeyabarrota*. Generally, it is in the *megeyabarrota* where the family sets up shelter for the duration of Torale.

By the beginning of May, the honeycombs all 'dry' due to excessive heat and the lack of rain. The end of honey signals that Torale is over and the rains will come soon. This temporal phase is called *obijakututante*, meaning 'we have eaten all that they could have had!'. The Ongees also regard the coming of *obijakututante* as a sign of success in creating *ikatallabe injube*, 'dryness within space'. Creating dryness in the space, eating all that the spirits could have had, causing the spirits to leave the island, and stopping the winds and rain are all achievements of the Ongees during Torale. They are all things that the Ongees do and the spirits suffer their impact. During the spirit-given seasons the inverse takes place: that is, the spirits return and control the winds, the rain, and food supply of the Ongees.

In the middle of May, a spirit called Dare (associated with

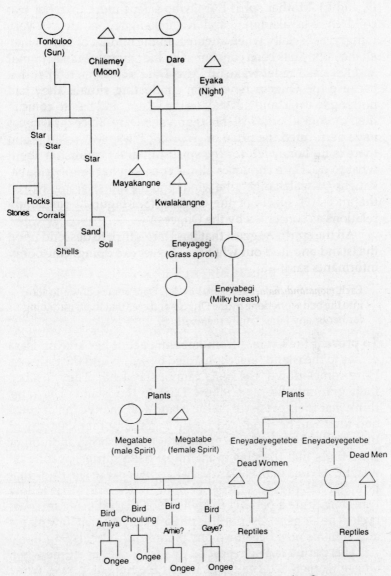

Fig. 2 Descent Relations as Conceived by the Ongees

the south-east winds) comes to the island. The spirit Dare is the mother of the spirit Kwalakangne? (south-west winds) and Mayakangne? (north-east winds) (*see* Myth no. 2, units A, B, C, D, and E). Mother spirit Dare, who sleeps more than her two children, Mayakangne? and Kwalakangne?, is always very hungry, especially when angered. In the middle of May, when all the honey has been consumed by the Ongees, Mayakangne? and Kwalakangne? wake up Dare. Dare's children tell her that because the Ongees have been conducting rituals, they had nothing to eat, could not stay on the island, and had to come to their mother's home. All this aggravates Dare. Since the Ongees have performed the ritual of *getankare*, the space is all dry and there is no honey left for the spirits. In great confusion about what to do, Dare convenes all the spirits in her family (Mayakangne?, Kwalakange?, the winds and the rains), along with all the ancestral spirits of the Ongees. (*See* Figure 2 for family relations as conceived by the Ongees.)

All the spirits suggest that it is Dare's turn to descend upon the island and find out what the Ongees are doing. One of my informants said,

> Each *monatandunamey* (seasonal cycle), Dare comes down to look into the bad work done by the Ongees and sees that there is nothing for her or any other family member!

To prevent the Ongees from knowing about her arrival, Dare comes to the island unaccompanied by winds and the rains. As Dare comes down, the rains cannot stay back. They disobey their grandmother and follow Dare to the island. The rains think that their presence on the island will change the dryness and food will be produced for their grandmother. Because of this disobedience of the granddaughters, the grandmother Dare finds that the pigs in the forest have started putting on weight. All this makes Dare very happy because she finds that now she has pigs to satisfy her hunger. The rains that follow Dare constitute a period of regeneration in the Ongee seasonal cycle. The Ongees describe this phase by saying that 'few plants and animals now become many plants and animals!'.

This period is also regarded by the Ongees as a time when Ongee women become pregnant. At the time of Torale, while all the spirits were away from the island, some spirits could not leave. The spirits remaining behind were trapped inside the honeycombs. During the Torale season, when women consume honey, spirits trapped in the honeycombs are released inside their wombs. The rains with Dare cause the spirit inside the women and the women themselves to put on weight.

During Dare everyone moves out of their respective shelters made in the forest. The new shelters are now set up near the creeks and the mangrove forest. The Ongees undertake this translocation so that angry and hungry Dare is not disturbed by their presence in the forest. Once the Ongees settle down in the region of the creeks and the mangrove forest, they start collecting and depending on fish, crabs, and fruits from the mangrove forest area.

By this time it is the month of June, and the Ongee are keeping watch on the wild pigs fattening in the forest and on the whirlwinds and spouts in the sea. Day after day, members of the pig hunters' band go to the forest to see how heavy and fat the pigs are. They will hunt one or two pigs and return to the camp at the creek set up by the turtle hunters' band. At the camp, the pig hunters present the pig to the turtle hunters and ask the question, 'Are they [the pigs] *idankuttuga* ['heavy'] enough?' The pigs are then cut and cooked. The skull of the pig is placed on a *kame* ('wooden scaffold') on the *batitujuney* ('horizon').

The *kame* (about five to eight feet high) are erected at the point where the coral reef ends and the seabed falls sharply away. It is at this juncture that the various spirits associated with the seasons arrive to visit the island. *Batitujuney* is a very significant spatial reference within the Ongee world-view. It is the point at which sky, land, and sea meet. The spirits coming down from various places always arrive at the island over the *kame* erected by the Ongees. The hunted pig's skull is placed on the *kame* for Dare to see and take note of the fact that the Ongees are killing pigs. Thus the *kame* are always erected and located in relation to the direction of the winds. For instance, the skulls to be seen and smelled by Dare are placed on a scaffold that is on the south-east horizon.

While the pig hunters continue to make additions to the pig skulls on the *kame*, the turtle hunters watch for whirlwinds and water spouts rising just beyond the *kame* on the south-eastern horizon. The dark water spout that connects the black clouds with the blue water of the sea is called *tegule*. When the *tegule* is sighted the turtle hunters come and report to the rest of their camp-mates:

> Dare is angry—she is really angry—she has changed into *Tenneya-bogalange* [angry and hungry form of Dare]. We have seen Dare going up the *tegule*. She has now gone to call the other spirits, winds, and rains—she knows that pig hunters have started killing the pigs that she was eating for herself.

After spotting the *tegule* ['departure of Dare'], for a few days the weather remains turbulent with occasional showers. On asking why Dare had left, I was told:

> Dare was hungry she came down to eat pigs. We had consumed all the honey during Torale. We Ongee, after eating honey, are all changed.[10] Pig hunters are now good at hearing (*gekalegebaro*) and the turtle hunters are now good at seeing (*gebogelah*). So we all hunt pigs even when Dare is here. This makes Dare very angry. Hungry and angry Dare has gone up *tegule*. She will then send down Kwalakangne? and Mayakangne?.

When the departure of Dare is announced by the turtle hunters, the Ongees proceed with the ritual of *tanageru* (initiation of young boys). The ritual of *tanageru* continues into late July. On the completion of *tanageru*, cyclones called *gingetigye*, with strong winds from the south-west, mark the arrival of Kwalakangne?. All the Ongees move from the creek and coasts to live in the shelters set up by the pig hunters in the forest. The coastal areas and the nearby mangrove forest, where the *tanageru* ritual was conducted, are now deserted. The strong winds on the coast make it impossible to face the sea waves, which rise up to about six or seven metres and cause much damage

---

[10] Evident in this statement is the special quality of the honey consumed by the Ongees. Not only is honey the spirit food and related to women's pregnancy, but its consumption by men makes them better hunters. Honey increases their quality of hearing in the forest, which aids them in locating and hunting pigs. In the same way, turtle hunters who depend much on the power of sight for the operation of turtle hunting improve their visual capacity by the consumption of honey. Thereby, pig hunters succeed in hunting pigs even in the presence of Dare in the forest and the turtle hunters are able to readily sight the departure of angry Dare. Above all, the whole community's consumption of honey enhances the hunters' powers and unites them to make Dare angry.

along the coastline. The translocation of the Ongees from the coastal area to the deeper forest is called *bey?tebe*. Once the groups move into the forest, pigs and tubers constitute the main diet. Kwalakangne? is a somewhat cold and wet season. These features are referred to as *garitabe*, which means to 'add weight'.

Around October the Ongees find it difficult to stay in the forest because of the increase in the daily rainfall. Subsiding winds allow them to move again towards the coastal areas. Winds from the south-west bring the Ongees out from the jungle to participate in the rarely successful dugong hunt.[11]

Myth No. 3

Dugong used to live in the forest. He called everyone to the follow-up session of singing and dancing. Everyone was invited but the *ayuge*, monitor lizard, and *kekele*, civet cat. Sitting on a tree, cat and lizard spotted all the pigs and the turtles going to the feast covered in red clay paint. Cat and lizard thought that though they had not been invited they should go. To do so they decided to steal some clay paint from the houses of ants and birds. White clay was brought from the house of ants. Cat went up the tree-tops and brought red clay from the house of birds. Cat and lizard decided to eat the two clays and go into the water for a dip. As a result, cat and lizard lost their body smell and changed colour. [Unit A.] Birds and ants found out about this incident and informed the pigs about it. However, the pigs did not know of this since the winds did not carry the smell of the cat and lizard, because they had consumed the clay paints.

[11] According to the Ongees, it is particularly difficult to hunt dugongs because of the way dugongs are related to both turtles and pigs. *Tuove*, dugongs, are animals which share an identity with the turtle as well as the pig. This shared identity makes them animals positioned between the forest and the sea. This idea is actually to be found in the Ongee myth of dugong.
Another myth tells that during the full moon tides, some pigs try to run away from the forest into the sea and some turtles try to run out of the sea into the forest. Because the pigs and turtles move in and out of water, the level of the sea changes and some animals are caught in-between. The animals caught between the forest and sea, in transit, become dugongs.
Therefore, the hunting of dugongs is rarely successful since they are difficult to locate. During my stay on the island I did spot some preserved skulls of dugongs. Only once did the Ongees return with dugongs while I was with them. Dugong is a delicacy for the Ongees and at some point in history, perhaps they were not so rare around Little Andamans, since Britishers did give the name Dugong Creek to the place where the majority of my research work was done.

Among the crowd of singers and dancers the pigs found out about the cat and lizard with the cooperation of birds and ants. Pig and lizard had a fight. They grappled for a long time. In the course of this fight the lizard dragged the pig to the water. It was in the sea that the lizard left the pig. It was in this throwing that the pig became dugong. All the birds were very angry and they decided to fight the cat and·lizard. [Unit B.]

The fight became a war. One by one all the birds were thrown into the sea and they changed into fish. Seeing all this turtle went and told this to crab. Crab went to lizard who was eating all the bird eggs and said that he was feeling cold and wanted some red clay. Lizard did not like his body— saying this lizard offered his nose into crab's claw. Crab took deep bite and cut off lizard's nose. Lizard in an utter state of pain and agony, fell down from the tree. All the ants collected and dragged lizard and threw him into the creek. [Unit C.]

After some days lizard recovered and became the crocodile in the creek. So cat lost her husband and in rage she started breaking trees. Spirits of monsoons did not like this and it rained and flooded the island. Everyone died in this flood. Some eggs were saved and out of that birds and fish were born again. The birds invited the turtle and dugong for a feast. Some turtles were able to go out and join the birds in the forest; they had much to eat and sing about. However at the full moon the dugong cried out on getting the smell of the turtles in the forest. Turtles did not like the idea of going back to sea, so the birds and ants gave the clay paints and the turtles changed into pigs and remained in the forests. [Unit D.]

By mid-October the Ongees have settled on the coast in the territory of the turtle hunters and are waiting for clear weather so that they may undertake turtle hunting. It is particularly unsafe for the turtle hunters to go out in the canoes that they have made in the seasons of Torale and Dare. The carved canoes have not yet dried and the sea has yet to calm down. In this phase of waiting, which is also the period of the shifting of the winds from south-west to north-east, tubers brought from the forest are boiled along with the crabs and small colourful coral reef fish. As the supply of tubers runs low, the Ongee women collect *daboja*, the fruit of the *Bruguiera cymnorihaza rizophoraceae*. *Daboja*, which grows in the mangrove forest, is carried down the creek to the sea and is the prime food for the turtles. The availability of *daboja* marks the readiness of Maya-kangne?, which is the turtle season. During this period Maya-kangne? (the spirit following Kwalakangne?) comes down to the island and satisfies its hunger by feeding on the pigs still in

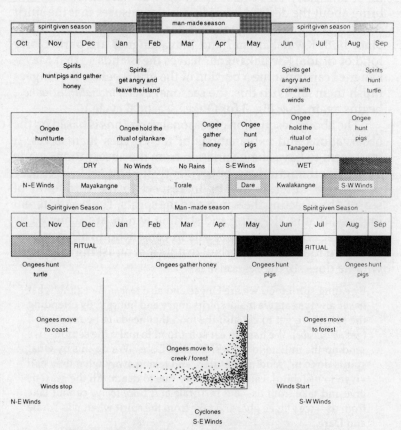

Fig. 3 *Monatandunamey*, Ongee Seasonal Cycle

the forest. The turtle hunters now get ready for the forthcoming turtle hunting season by giving finishing touches to their canoes. They also cover the fibrous ropes, which are used for tying the turtle hunting harpoons, with beeswax. To make certain that Kwalakangne? is gone and Mayakangne? has arrived, the men go to the forest and collect *tombowage*, 'cicada grubs'. These are just like honey (derived from insects, a spirit food, not requiring any mastication, and associated with the coming of Dare), and the Ongee camp-mates eat cicada for a specific desired effect. The consumption of cicada in late Oc-

tober and early November is thought to start the winds and
bring about the desired humidity. It also ensures that the spirit
of Kwalakangne? has gone and Mayakangne? has returned.
Since both Kwalakangne? and Mayakangne? are particularly
fond of cicada, Kwalakangne? leaves the islands so that Mayakangne? can consume a portion of the grubs before the Ongees
finish them all. With this, the seasonal cycle, *monatandunamey*,
begins again. (*See* Fig. 3 for Ongee seasonal cycle.)

The above description of seasonal variations is based on the
observations and explanations of the Ongees themselves. It
may seem to have some variation from the observations made
by earlier ethnographers, but evidently the Ongees believe they
obtain two different results in respect to the formation of
seasons by effecting the winds and the spirits. The two different
impacts, that is, stopping and starting winds, are achieved by
the Ongees offending the spirits in conjunction with consuming 'spirit food'. It is important to note that on asking the Ongee
why he does so, the response is:

> By eating spirit food we are Ongee and also *tomya*. By eating what
> *tomya* always eats we make spirits angry and hungry. By offending
> the spirits we get to eat also the food that needs to be masticated
> (*geetakwawabe*). We have to eat spirit food to make the season—we
> send up the initiate to the residence of Dare—We do this by what
> spirits do to us, sending children, and taking away what they had
> given to us [child], causing us to cry due to death. All the seasons
> give and take from us—but in Torale and Dare *tomya* cannot take
> from us—we Ongee give and take from the spirit when it is Torale
> and Dare.

CHAPTER 2

Origin of Questions

Radcliffe-Brown and the Question of Power in
Andamanese Culture

Between 1906 and 1908, A.R.Radcliffe-Brown conducted field-work for his classic ethnographic study, *The Andaman Islanders* ([1922], 1964). He documented in detail the social organization of widespread groups of the islanders, their ceremonial customs, religious and magical beliefs, and myths and legends. Radcliffe-Brown was able to collect numerous variants of Andamanese myths and rituals because the essential ideas of the myths and rituals were shared by several dialect groups then living on the Andaman Islands. Today the islands are vastly different and only four dialect groups remain. Radcliffe-Brown's work, of course, forms a major background for my field-work and formulation of ideas and analysis.

   In considering the Andamanese system of ritual and myth, Radcliffe-Brown develops a hypothesis about the relationship between ritual and myth within the larger context of Andamanese religion. He argues that, for the Andamanese, religion has two important aspects: a belief in nature, and an organized relationship of power between man and higher powers (Radcliffe-Brown, 1964: 405). This power relationship has a moral character, and is one of the principles that regulates and organizes the islanders' ceremonial life, so that harmony is established and maintained. The fundamental premise of this formulation is derived from Radcliffe-Brown's notion of an intrinsic opposition between man and spirit (Radcliffe-Brown, 1964: 307). His analysis emphasizes the social value of power, ceremony, natural phenomena, and religion. Hence, when he discusses the seasonal winds that generate the various seasons and weather conditions, he suggests that the,

Andaman Islanders express the social value of the phenomena of the weather and the season, i.e.the way these phenomena affect the social life and the sentiments, by means of legends and beliefs relating to the two mythical beings . . . we may say that the Andamanese personify weather and the seasons [Radcliffe-Brown, 1964: 353].

He also states that

Between the rainy season proper and the cool season is a period of six or eight weeks in which the weather is unsettled; the wind is variable; fine weather alternates with storms that are sometimes of terrific violence; waterspouts are frequent; it is in this season that violent cyclonic storms are likely to occur. This season is called by the Andamanese Kimil . . . the word *kimil* denotes a condition of social danger, or of contact with the power possessed by all things that can affect the life and safety of the society . . . The life of the Andaman Islander is profoundly affected by the alternation of the seasons. [Radcliffe-Brown, 1964: 352.]

For Radcliffe-Brown, anything capable of affecting social life has social value, and the winds are no exception to this. As the winds shift, the seasons change, and this in turn causes a change in the islanders' source of food. In response to the seasonal changes, the islanders (including the Ongees) follow a translocationary cycle. The forest dwellers are hunters of pigs and the coastal dwellers are hunters of turtle. In search of food, each group regularly moves beyond its particular traditional resource base. During the dry season, when pig hunting is proscribed, pig hunters move from the forest to the coast and are supported by the turtle hunters. In the rainy season, storms make turtle hunting impossible. Consequently, turtle hunters move to the forest and are supported by the pig hunters. The basis of this cycle is the notion of unfavourable weather conditions, and Radcliffe-Brown explains the 'bad weather' as being the result of contact between two opposites, human beings and spirits (Radcliffe-Brown, 1964: 163). When the spirits are offended by their contact with man, they express their anger either by sending bad weather or by leaving the island. Some spirits are associated with the sea and the forest, where they reside, while other spirits, such as Biliku and Tarai, are associated with various phenomena of nature.

Thus the islanders move from place to place in search of

food and to escape the strong winds unleashed by the anger of the spirits. The movement of humans is interdependent with and reactive to the mood of the spirits. The immediate question, as Radcliffe-Brown points out (1964: 356), is why, given the consequences, do the islanders offend the spirits who are related to the winds?

Men and Spirits: Relations of Anger and Offence

Radcliffe-Brown points out that the Andaman Islanders do all those things that they are not supposed to do, thus making the spirits angry and causing storms (Radcliffe-Brown, 1964: 152-3). Ethnographic accounts also point out that, at the same time, efforts are made by the islanders to control harsh weather. For instance, the most efficacious means of controlling weather (chiefly, stopping storms) is to destroy things such as beeswax and certain creepers (Radcliffe-Brown, 1964: 157), which are specifically regarded as belonging to spirits like Biliku. But the destruction of these materials angers the spirits to whom they belong. First, the tribesmen make the spirits angry and cause storms (Radcliffe-Brown, 1964: 153-4). Second, doing the same things to the same substances (Radcliffe-Brown, 1964: 157) is regarded as 'heroic remedies' against weather conditions associated with the anger of the spirits. This raises questions about how and why Andaman Islanders deal with the destruction and consumption of certain items that may have two different or even contradictory results.

> The explanation that I have to offer of these beliefs relating to Biliku and to the things that offend her is that they are simply the statement in a special form of observable facts of nature. The rainy season comes to an end, the wind becomes variable, yams and other vegetable products begin to ripen and are used for food, and stormy weather comes, some years bringing cyclones of exceptional violence. Then follows a period of steady N.E. winds with fine weather and abundance of vegetable foods, during which the noise of the cicada is not to be heard. Then comes the honey season, when everyone is busy collecting honey and melting bees'-wax. The wind becomes very variable, storms come, the fine weather comes to an end and the rainy season begins again. These facts affect the feelings of the Andaman Islander and he expresses his impressions by

regarding all these happenings as if they were the actions of an anthropomorphic being. The vegetable products, the cicada, and the honey all belong to Biliku. When the yams are dug up she is angry, or in other words, storms occur; a storm *is* the anger of Biliku. The cessation of the song of the cicada removes one of the possible causes of the anger of Biliku, and therefore marks the period of fine weather. That anger appears once more when the natives busy themselves with melting bees'-wax.

It may be noted that these beliefs about Biliku give an expression of the social value of honey and bees'-wax and of vegetable foods such as yams. The Andaman Islands provide few fruits containing natural sugar. Yet the natives are inordinately fond of sweet things; they greatly enjoy the sugar that they now obtain from the Settlement of Port Blair. Honey, which was almost their only sweet food in former times, was therefore very greatly valued. Apart from the yams and other foods associated with Biliku there are very few productions of the Andamans containing starch in a palatable form. To the native who has been living during the rainy season almost entirely on meat and fish, the starchy foods of the stormy season (yams, *Caryota*, etc.) are of great value, and they are very highly prized. Thus the foods associated with Biliku all have a high value.

We all know how the value of an object is increased if, in order to obtain it, we have to make some considerable effort or sacrifice, or put ourselves in danger of some evil. Reversing this mental process, the Andaman Islander expresses his sense of the value of honey and yams by the statement that to obtain them he must be prepared to risk the anger of Biliku with its results.

[Radcliffe-Brown, 1964: 362-3.]

Radcliffe-Brown's explanation of the storm as the anger of the spirits poses a problem. How can the destruction and consumption of the substances have two different effects, that is, first, to stop the winds and pacify the angry spirits, and second, to anger the spirits and start the storm? In my approach to this problem and its analysis, I attach significance to Radcliffe-Brown's suggestion that in order to obtain something, one has to risk the anger of the spirits, and that, consequently, to make the spirits angry makes it possible for the islanders to obtain specific items of food (Radcliffe-Brown, 1964: 364).

Radcliffe-Brown's ethnometeorological accounts reveal that the spirits have the power to affect humans in the form of weather and that, similarly, the people on the islands have the power either to offend or to pacify the spirits and hence to control the weather. Clearly, people and spirits interact in a complex relationship of power. The root of power and of the power relations between the islanders and the spirits lies in the different impacts of the same acts. Humans and spirits are not just opposed to each other but rather exist in a dynamic relationship. There are reciprocal actions and reactive forces between the powerful and the powerless. Thus, what is conditionally available to humans because it is due to the spirits, is made available by humans to themselves through the conditioning of the spirits. Storms and the anger of the spirits are the conditions and impacts of this kind that are shared between humans and spirits.

However, in Radcliffe-Brown's formulation, power inheres in all objects or beings that in any way affect social life for good or bad. Further, power is the basis of society on which the members of that society depend for their well-being. In relation to human society, spirits are dangerous. In relation to the powerful position of the spirits, human individuals and the society constituted by them perceive their protection in the collective form (Radcliffe-Brown, 1964: 307). If Radcliffe-Brown's perception is right, then do the islanders also condition the ways in which power becomes available through the acts and impacts that relate people and spirits?

Radcliffe-Brown's presentation is built on a fundamental paradox: the very acts that offend the spirits and generate bad weather also placate the spirits and control the weather. His explanation is less a resolution of the contradiction than a description of a multi-tiered system of equilibrium: contact with power is dangerous, but the risk is avoided by means of ritual precautions. For example, the 'socially valuable' and 'magically potent' fire, bones, and clay paints are all used to protect the tribesman in his various encounters with spirits (Radcliffe-Brown, 1964: 184, 264). These items have magical potency and social value because they have power. In describing the nature of that power, Radcliffe-Brown argues that the degree of power possessed by anything is directly proportional to the importance of its effect on social life. However, even as a structural–functionalist argument, Radcliffe-Brown's reasoning is subject to important qualifications: the islanders cannot avoid offending the spirits, given the seasonal cycle and change in relation to the resources available to the former. Indeed, there is no account of any ritual precaution in Radcliffe-Brown's ethnography to support his formulations about power, especially about the power relationships between the islanders and the spirits. On the contrary, his ethnographic account and analysis draw repetitive parallels: the same offensive acts enable the islanders to restrict and restrain the anger of the spirits and to stop the storms.

Although the Andamanese experience inescapable spirit-induced oscillations in weather conditions, food supply, and areas of residence, there are individuals within the community who are able to stop storms (Radcliffe-Brown, 1964: 157). This leads us to a point of assumption that at certain times certain individuals have a power equal to or even greater in potency than that of the spirits. The conclusion is that power, its acquisition, the relations formed on its basis, and the character of its possessors must be of some importance to Andamanese culture itself and to its study.

Some details of Radcliffe-Brown's analysis of power and its ritual embodiment are necessary here. Objects and beings are embodied with power if they have some effect on the social life of the community. This effect may produce good or bad results, it may aid or harm the society. This presupposition (Radcliffe-Brown, 1964: 307) forms the foundation for Radcliffe-Brown's central thesis of the opposition between man and spirit. This formulation is correct so far as it goes; but it does have two fundamental weaknesses: (a) the power relations between man and spirit are not in static equilibrium, and (b), an adequate analysis of the ritual endowment of that power must go beyond Radcliffe-Brown's notion of ritual as a meta-social projective system capable of converting social sentiments and values into communications.

One of the important and interesting ideas developed by Radcliffe-Brown in his account of the Andaman Islanders is:

> If an individual comes into contact with the power in any thing and successfully avoids the danger of such contact, he becomes himself endowed with power of the same kind as that with which he is in contact.
>
> [Radcliffe-Brown, 1964: 306.]

In light of this formulation it is worthwhile to consider the rituals within Andamanese society to find out if they provide any situation and/or context of 'contact with power'. If they do, then what is the relation between man and spirit? It may not be static, since Radcliffe-Brown does tell us of individuals who are in contact with the spirits, i.e. Oku-Jummu and Oko-Paid ('Dreamer and Medicine Man') (Radcliffe-Brown, 1964: 176). Does contact with spirits who have the power to generate storms give power to the individuals to stop storms? If it does, then the individual who has established contact with spirits has gained power and, in stopping the storms, demonstrates his sharing of some identity with the spirits. Therefore, it is valid to question the power relation between man and spirit described by Radcliffe-Brown. That is to say, power is not confined either to individuals or to spirits. Individual human beings can contact the spirits to gain power and use it. This makes the aspects of power capable of being transacted between humans and spirits. Since power itself can be transacted, the relation between humans and spirit by nature is characteristically dynamic. The relations of power between humans and spirits are capable of shifting and changing. Since this relation of power is non-static, humans and spirits are related to each other in more ways than that described by Radcliffe-Brown, where people belong to the 'world of the living' and the spirits belong to the world of the 'dead' (Radcliffe-Brown, 1964: 307).

Certain limitations of Radcliffe-Brown's theory of ritual have been noted by earlier scholars. From a structuralist perspective, E.Leach (1971: 40-1) criticizes Radcliffe-Brown's tendency to isolate myths and rituals as discrete symbolic units. Instead, Leach focuses on them as a class of symbolic incidents affecting a transformational process, so that the controlling of storms or the offensive acts generating storms are seen to be agents of transformation. Despite this promising beginning, Leach does not explain how or why the same symbolic act has two different impacts, nor does he attempt to trace the transformation of an agent in the process of change or in the examined corpus of Andamanese myths. Leach contends that transformation has its converse in 'real' life (1971: 36), from Levi-Strauss's position that the basic preoccupation of myth is the ambiguous borderline between what is natural and what is cultural (*see* Claude Levi-Strauss, 'The Culinary Triangle', *Partisan Review* 33 [1966]). I do not suggest that Levi-Strauss's is a wrong formulation, specifically in light of the myths recorded by Radcliffe-Brown. Leach's analysis does, however, underscore, for example, the great relevance of his formulation about myths being an ambiguous borderline between Nature and Culture. Let us consider a case within the Andamanese context.

In ancient times, fire was under the exclusive control of Biliku who gave fire to 'Sir Prawn'. The kingfisher stole fire to roast, he ate and then went to sleep. Then the dove stole the fire and gave it to the Andamanese who were taught by Biliku to cook the food (*from* E.H. Man, 1885).

Kingfisher's theft was punished when Biliku cut off the bird's head and transformed his status from that of an ancestor to a mere bird. Thereafter, kingfisher could eat only raw food; the Andamanese could cook, but lost the ability to fly. Fire, then, transforms raw food to cooked food, and distinguishes birds from men. We should note that the Andamanese consider roasting more dangerous than boiling; the roasted food served on a ritual occasion marks exocuisine; the Andamanese otherwise prefer boiled food, endocuisine. Thus, roasted food is natural and boiled is cultural (Levi-Strauss, 1966); and both are distinguished by the use of fire. Fire also distinguishes, by its possession, humans and non-humans, i.e. birds. Those ancestors who were afraid of, or were burnt by, fire became birds, beasts, and fish—cut off from human society which 'from that moment constitutes itself around fire' (Radcliffe-Brown, 1964: 342).

Kingfisher's stolen fire enabled him to cook; but once the fire was lost he had to eat raw food and assume a natural identity. Fire, then differentiates men from animals, and nature from culture; its absence destroys the nature/culture distinction. Andamanese mythology, thereby, presents events in which the acquisition of fire, a power, creates a distinction between men (culture) and animals (nature). The absence of this power again switches nature and culture: men without fire are like animals, and vice versa.

Kingfisher exemplifies fire mythologic: by stealing fire he is like men, but having lost its fire he is just a bird. The implication is that the possession of power constitutes one as a cultural entity, which again presupposes that to gain power one must start as a natural entity.

There is need to reconsider the power of ritual in Andamanese culture by contrasting Radcliffe-Brown's rich data with his rather unsatisfactory analysis of it. Leach argues that 'the general effect of ritual must be to present the participants with a conception of the cosmos' (1971: 42).

Leach's criticism of Radcliffe-Brown's theory of ritual has suggested important questions for my own research. Do the Andaman Islanders still live in a state similar to that described by E.H.Man (1885) and Radcliffe-Brown (1922)? Are their rituals based on a 'conception of the cosmos?' What do the changing power relations among humans, animals, and spirits indicate about the character of power and its possessors, and the transformation of nature and culture? Does the ritual act, which can generate or stop storms and winds, also exchange or switch the identities of the offended (spirits) and the offender (humans)? Do these rituals track the seasons and indicate a cycle of 'natural' seasons generated by the spirits and 'cultural' seasons effected by people? If so, are the living hunters and food gatherers of the Andaman Islands aware of the cosmos, and do they deliberately attempt to gain power over people? How do men gain power and, in the process of gaining it, what is their identity?

In my examination of these questions and observations made through a study of Ongee culture and ritual, I posit that in the process of gaining power over or from the spirits, humans not only appropriate power from the spirits but also assert their identity with spirits.

The anger of the spirits, whether self-induced or caused by the actions of humans, has an effect on the islanders' safety, on the availability of resources, and on social existence at large. Since humans and spirits are equally capable of starting and stopping storms and of affecting each other, the homology of acts and impacts between humans and spirits shows that the identities of powerful spirits and powerless humans are subject to shifts and reversals. By making the spirits angry, men gain things such as food, and become like the spirits themselves, since the spirits and their anger are determinants of what is available to people. When the spirits get angry, they leave the island and the food consumed or processed by them becomes available to people. This sets up the possibility of an identity between humans and spirits, since the consumption of a food resource, either by humans or by the spirits, makes it possible for only one of the two to consume it. Just as the spirits condition men's capacity to consume resources, humans also effect the spirits' capacity to consume.

The available ethnographic records have little to say about the Andamanese perspective on rituals. Radcliffe-Brown's *The Andaman Islanders* (1922,1964), the most complete work extant, excludes such information because of his theoretical position, and because the society was already in an advanced stage of population decline and socio-cultural disintegration. E.H. Man's study on the *Andamanese* (1885), based on his observations as an administrative officer from 1869 to 1880, contains important data, but is so general as to be of little use in a serious study of ritual. In 1952, Lidio Cipriani lived for 162 days among the Ongees of Little Andamans. His monograph, published posthumously in 1966, is weak in ethnographic detail but does illustrate a certain continuity of life-style from the period of E.H. Man and A.R. Radcliffe-Brown. Finally, the work of the physical anthropologist P.Ganguly (1961) suggests an elaborate 'conception of the cosmos' and ritual life in Andamanese society. Oblivious to rich ideas and clues present in the works of Radcliffe-Brown (1922) and Ganguly (1961), a monograph on the Ongees published in 1990 (Basu, 1990) represent Ongees as 'The Onge have a few rites and rituals . . . In fact they do not have an organized religion as such, but rather some beliefs, and fears emanating from them' (Basu, 1990: 66).

Power of Humans, Spirits, Winds, and Rituals of Movement Among the Ongees

Based on Radcliffe-Brown's account, my prime concern in the field was to understand how the Andaman Islanders elicit ideas about the impact of power between humans and spirits. From the viewpoint of the ritual participants, rituals can alter the state of the world because they invoke power.[1] If power is

---

[1] This kind of approach to power predominates ethnographic accounts and an-

treated as inherent in the ritual itself, action by the ritual participant is called magic. If power is believed to be external to the situation, then the analyst understands it as a supernatural agency and religion (Schmidt, 1910). Arguments on this theme have been highly contentious, but my position, formulated through an understanding of Andamanese culture, is that ritual or rite is prior to the explanatory belief, a view ascribed in anthropological circles essentially to Robertson Smith (1956).

The concept of power itself is a derivation. For example, A asserts dominance over another individual, B; we observe that B submits to A, and we say that 'A has power over B'. In the context of ritual one can observe another individual, A1, going through a performance that he believes will coerce a fourth individual, B1; or alternatively, we can observe B1 making a ritual act of submission to an unseen presence, C1. Normally one would classify and declare that the acts of A and B are rational but that the acts of A1 and B1 are irrational. For the Ongees the actions of A, B, A1, and B1 are all of the same kind. The 'authority' by which A is able to coerce and control the behaviour of B in a secular situation is just as abstract and metaphysical as the magical power by which A1 seeks to coerce B1 or the religious power that B1 seeks to draw from C1.

Relations and ideas about relations with spirits and human beings or about the potency of a particular ritual are modelled

---

thropological analysis of *mana*. For example, Codrington, in discussing Melanesia in 1891 (1965:256), said that *mana* accomplishes everything beyond the ordinary power of men. If people and objects have it, its source is the spirits. Lessa and Vogt (1965), on the other hand, explain *mana* as referring to 'sheer power—occult force independent of either persons or spirits'. *Mana* for them is dangerous and restrictions are necessary; so wherever there is *mana* there is a *tapu* ('prohibition for protection') (Lesa and Vogt, 1965:253). Raymond Firth's analysis of 1940 had deplored the abstract discussions showing how Polynesian Tikopians use the Polynesian word for concrete situations and concrete results only. *Mana* does not exist in a vacuum; it is always *mana* of a person or thing. Tikopians insist that it comes from the gods. It is used for human benefit and it may refer to the cure of sickness (Firth, 1967:183, 189, 192). Hocart's works on Fiji elaborate on *mana* as medicine for ghosts and spirits and for chiefs whose curses come true. Some medicines in Fiji are made effective through spirits, and the leaves of some trees have *mana* only in the hands of certain men. Hocart gives a synonym for *mana* in medicine that translates as 'hits the mark'. Hocart also gives an example of the use of *mana* in Tongan as a curse that is effective (Hocart, 1914:98-9)

on first-hand experience of real life relationships formed
around the act of hunting, a life-taking and smell-manipulating
activity. Conversely, every act by which an individual asserts
his authority to curb or alter the behaviour of another individual, as is the case between humankind and spirits by the
use of *gobolagnane*, is an invocation of a metaphysical force.
Individuals are influenced by magical performances or religious implications, just as they are influenced by the commands
of authority. The power of ritual is just as actual as the power
of command (Tambiah, 1979).

Among Ongees, individuals who contact and communi-
cate with the spirits to gain special knowledge are called *torale*
and are distinct from those who avoid all possible contact with
the spirits. Contact with the spirits (*enegeteebe*) is to be avoided
but is always possible, since humans and spirits share the
island's forest, sea, and sky. The various places that constitute
space are all subject to change, because through them and in
them various spirits move along with the winds. In relation to
the movements of the spirits and the translocations of the
islanders in time and space, aspects of Ongee cosmology are
shared by and between the Ongees and the spirits (*tomya*). The
elements of Ongee cosmology such as time and space and
related movements of humans and the spirits generate acts that
have impacts on not only each other but also on Ongee cosmology. Ongees use certain 'magically potent substances' (Radcliffe-Brown, 1964: 184, 264) to deal with the impact of spirit
contact. Within the space and the places that constitute it, spirit
movement has negative and positive impacts on the Ongees.
Since all living things are capable of releasing smell, the Ongees
use clay paints, fire, and preserved bones to restrict or release
smell, which, when carried by the winds, has the desired effect
on the spirits being attracted or warded off from the source of
the smell. In other words, the Ongees use the paints, fire, and
bones to effect the movements of the winds. Thus, by using the
'magical substances', a counter step taken by the Ongees, it
becomes possible to have an impact on the movement of the
spirits. However, the case of the *torale* is different. The *torale*,
who does not use the 'magical substances', foregoes certain
measures of safety and puts himself in a dangerous position,
so that he can undertake the special movement with the spirits

to their place of residence. When the *torale* goes with the spirits, which could become his death, a special knowledge is gained by the whole Ongee community, provided that the *torale* comes back from the spirits' home. Identical to the context of the *torale*, Ongees offend the spirits, put themselves in a dangerous position, and use that position to manipulate power to initiate the young males of their society. The ritual, using the grammar of movements, smell, and 'magical substances', actually enables the Ongees to do what the spirits do throughout the seasonal cycle, that is, to give birth and death. Through the ritual of initiation (*tanageru*), the Ongees cause the death and birth of a young man, which, in all other circumstances, are caused by the spirits. Before the Ongees can do this, however, they offend the spirits so that the angry spirits leave the island, and thus create a situation in which people act, and the spirits experience the impact of those actions. This forms a special durational division within the Ongee seasonal cycle. This interval, as in Radcliffe-Brown's accounts, is marked by humans stopping the winds and then starting the winds again. Once the initiation of the young men, which like the case of the *torale* involves undertaking a movement to the spirits' home, is over (i.e. the initiate returns), the relation of power between man and spirit changes back again. Prior to and after the ritual the spirits act and the Ongees experience the impact of those actions. During the ritual, the Ongees act and the spirits experience the impact. Through a consideration of these shifts and switches of act and impact between humans and spirits, what is established by the Ongee culture for an Ongee is that, in order to become powerful and to gain power, those who are already powerful are to be made powerless. This is not just a give and take of power. Rather, the power of the spirits and of humans gets externalized and objectified in magical substances, such as the *torale* and the initiate. This does not happen for a prolonged period of time but for a short duration during a movement, so that the self-reciprocating asymmetrical power-relation between humans and spirits is sustained. The significance of space, of moving in it, and sharing it with the spirits throughout the whole seasonal cycle may be seen in the pervasive role of socially valuable and magically potent objects, such as fire, smoke, clay paints, and bones (Radcliffe-Brown, 1964: 184, 264).

This awareness is a fundamental part of Ongee cosmology (cf. Leach's notion of cosmos, 1971: 42).

From the perspective of the Ongees, 'cosmology' is a cultural construct of space, which is an aggregate of places such as the sea, the forest, and the sky. All the places taken together constitute space (*injube*). Each place is either the residence of a specific spirit or is capable of being visited by a specific spirit. Human beings also reside in and move through places, and consequently humans and spirits are co-residents of *injube*. Smell and the winds also move within *injube*, and they alone can move with a capacity beyond that of human movement. The human and animal capacity for movement is perceived by the Ongees in relation to the movement capacity of spirits. The spirits (*tomya*) move (*eyolobe*) frequently between places (*nanchugey*), but it is concurrent movement along the vertical and horizontal planes (*oatatekoyabe*) that is of particular relevance to the coincidence of human and spirit movement. The movement of the wind (*tototey*) is also multidirectional and interconnective. Indeed, the Ongees have a special term (*gakweoney?*) to denote wind movement that interconnects the vertical and horizontal axes and transfers smell (the monitor of spirit movement) from place to place. It is this aspect that makes anemology one of the key analytical concepts in ethnographic representation of Ongee culture. The Ongees themselves, however, move on either one axis or the other, not on both. The terms for horizontal movement (*geeroyebe*) or vertical movement (*wabekomabe*) are assigned only to physically animate beings. As the Ongees move, they are constantly aware of the translocationary flux within the cosmos and utilize various magically potent (i.e. socially valuable) objects[2] to avoid the spirits' path. The Ongee moves in a cartographically 'safe' way. This safety is essential for providing the constant movement necessary to follow the regular cycle of food sources. Contact with a spirit (*enegeteebe*) under such circumstances is not necessarily bad (results can include loss of tools or even of life), but must be controlled. Occasionally ancestral spirits assist their living relatives in locating food or other valuables;

---

[2] Magically potent objects, such as fire, smoke, clay paints, and bones, control the release of the smell and odours which attract or repel the spirits.

such positive encounters (*talabuka*) are the deliberate conjunctions in space of man and spirit. The negative counterpart
(*malabuka*) implies coincidence and it is an encounter that occurs unexpectedly.

The importance of the movement of humans and spirits is
complemented in Andamanese ritual by that of the movement
of the wind and seasons (*see* Figure 3), a rough sketch of which
follows:

Mayakangne?	October–February: rain and north- east winds
Torale	April–May: dry; no winds and no rains
Dare	Late May–July: south-east winds
Kwalakangne?	August–October: south-west winds

In each of the above terms (with the exception of Torale) the
season's name simultaneously denotes the corresponding
spirit, the related wind (considered the spirit's 'grandchild'),
and the season itself.

The four divisions of the Ongee seasonal cycle are in themselves evidence for the formulation that the power of ritual lies
in a change of identity, and that the power relations of people
and spirits are subject to shifts and switches due to the ideas of
act and impact along with coincidence and conjunctions. Mayakangne? and Kwalakangne? are regarded by the Ongees as
seasons experienced as the voluntary arrival and departure of
the spirits and are therefore perceived as 'spirit-given seasons'.
The seasons of Torale and Dare differ from Mayakangne? and
Kwalakangne?. Torale and Dare together form a duration within the Ongee seasonal cycle during which humans act and
spirits receive the impact of those acts. During Torale, the
Ongees conduct an offensive ritual (*beti*), known as *getankare*,
by means of which the rains and winds are stopped. The
Ongees consider the stopping of wind and rain to be caused by
humans when they expel the spirits by consuming honey, the
prime food of the spirits, which is proscribed for humankind.
From April to May, the Ongees gather all the honey they can,
which results in the angry spirits leaving the island. As the
spirits depart from the island, the Ongees undertake the initiation ritual (from late May to July). During this period, an

excessive number of pigs are hunted (also proscribed by the spirits for this duration of the seasonal cycle) and the initiate is 'sent up to visit the spirits'. This 'sending up' of the initiate to the spirits during the initiation ritual, and the subsequent activity of 'bringing him down' is not only a distinct moment within Ongee culture, but also presents a situation unique in relation to all other circumstances. In the course of day-to-day life, the spirits on their own carry out the acts of moving about and of taking away the living Ongees. Spirits are also respon-sible on their own for coming back and bringing back the Ongees. Within the Ongee world-view, the successful return of the initiate from the spirits depends on the way the tribe conducts the ritual of initiation and, above all else, on how it brings back the spirits to the island on which they reside. The basic pattern of movement by the initiate who is sent by the community to the residence of the spirits is identical with the movement to the spirits undertaken by the *torale* on behalf of the community except that the *torale* accomplishes his journey alone and unaided by the community. It is here that the con-cepts of spirits, danger, rituals, and movements coalesce to explain Ongee notions of the power of ritual.

For the Ongee, ritual involves not only the formation of social identity, but also invests the Ongee with power; in-dividually by the bestowal of 'hunting' prowess, and collec-tively in the manipulation of seasonal change. It is because of this that one of the *torale* said, 'To live like an Ongee is to hunt everything in relation to the spirits. If not, then we Ongees will be hunted by the spirits'. Thus the Ongees, by realizing that they too can be hunted while they hunt, firmly establish a belief in the cause and effect relationship between the spirits and themselves, and a belief that rituals, such as that of initiation, make individuals like the 'magical potent' substances, such as clay, fire, and bones. The object of Ongee ritual is to give an individual body a form in which it can act and have an impact like the 'substance'. In doing this, the individual Ongee exer-cises the same impact as the 'substances' have on the spirits. Shifting and switching identities and the means of exercising and exchanging form the core of Ongee cultural praxis.

On reaching the forest home of the Ongees, a question that was continuously on my mind was: When will this camp move from here to another place? It was the month of October 1983, and I had been living at the coastal camp near Dugong Creek for a month. I was in a place where, to our east, was the Andaman sea, and on the west was the deep tropical rain-forest with enormous trees, in many cases reaching over one hundred feet. The tree-tops form a canopy that cuts out all direct sunlight on the ground. Although the trees are so large and heavy, their roots tend to be shallow and they are easily toppled over in the heavy storms that frequent the Little Andamans. The Ongees were aware of all this, so they were waiting for the wind conditions to change before moving into the forest. I was tired of eating fish and turtle meat and wondered when I would be moving with the Ongees into the thick tropical forest, which was so overwhelming and appealing from the white sands of Dugong Creek. For the Ongees, living on the coast was a safe place, but for me the coastal camp was just one place among others that I wanted to visit with them. My city background made me eager to enter the forest, but a sense of fear overpowered curiosity and kept me tied to the then strange campmates with whom I was living. To move from place to place in a given space for me was the essence of freedom. The associations of place with safety and security, and of space with freedom and exploration, were my first realizations about our culture, while living with the Ongees at Dugong Creek. In due course, I learned the *ceye?ne* language of the Ongees. In *ceye?ne* the use of prefixes and suffixes makes it possible to trace meanings and ideas in each word (Bloch, 1949; Radcliffe-Brown, 1914). For example, *ko* means 'where' and *ale* means 'children'. By putting together ko-r-ale the term becomes *korale*, 'home, a place where children are'.

As I started acquiring a better grasp of the language, I started posing the question to all the people around me: When will we move from here to another place? I never got any specific answer in terms of time. There was only one answer, and it was rhetorical: 'This place is good—when the forest is good we move into forest place—then the sea place is bad!'

What makes a place good and bad or safe and dangerous, and when do settlement and movement make sense? This was the first question in my notebook. Besides, I also had to find out if the concerns in my mind were questions for the Ongee in his own cultural context. It is important to understand how different cultures would and do answer questions. However, the questions remain questions only for anthropologists aligned with the school of ethnosociology. Is the anthropologists' 'Notes and Queries' at all a question for the people in a given culture? In order to convey an ethnographic account which is founded on the culture's own articulation of its own concerns, this account of Ongee culture is structured around the questions that are important for the Ongees as well as significant to Andamanese ethnography.

I learned that how and when the Ongees move depend on all that moves within the Ongee space. Just as I was interested in mapping out the movements of the Ongee hunters and gatherers, the islanders were on the look out for what moved around them and what they could do in relation to the various aspects of movement and to the elements that moved. For the Ongees, the elements referred not only to the natural aspects of Ongee space, such as the tides, sun, moon, clouds, earthquakes, and winds, but also to the animals, such as pigs, turtles, dugongs, monitor-lizards, civet cats, and insects, as well as to the spirits. These elements constitute an integral part of the Ongee world-view. The places in which the elements move are interrelated by the relation of act and impact within the whole space. Every element's movement, an act, has an impact on other elements. This combination of an element's act–impact relation sets up a series of transformations. This particular viewpoint makes the Ongee cosmology a stage of movements where space and place are never empty. The awareness of different elements and their movements within different places, guides Ongee action. Consequently, the cosmology in which Ongees, elements, places, and movement are located acquires an ideological value. The ideological value and the cartographical implications of Ongee cosmology organizes the world of the Ongees in which social processes are effective.

The Ongee cosmology, however, is not based on just the idea of nature, but also on how nature is related to the social. The recognition of this relationship between nature, where things move, and society, where things are accordingly moved, leads to the emergence of various values as cultural reality. Equality and hierarchy of humans and spirits, things and ideas, are therefore embedded in the context of Ongee cosmology, and are signifiers and signified through the various forms of movement of various elements. Spirits, human beings, and animals are distinguished from one another on the basis of their differing moving capacities and body conditions. They are in a hierarchical relationship with one another in which a body that has smell and a body that does not have smell are 'encompassed' elements. Bodies that move on both the vertical and horizontal axes and bodies that move only on one axis or the other are the 'encompassing' features of the hierarchical system.

In order to understand Ongee cosmology as a 'structure' in the sense that L.Dumont (1970) uses it, the fundamental opposition between humans and spirits and between humans and animals must be located. This opposition is based on the possession of smell: bodies without smell move along both the vertical and horizontal axes of Ongee space. The loss of smell or the acquisition of smell changes the position of bodies within the hierarchical structure. The whole of Ongee cosmology is founded on the necessary and hierarchical coexistence of spirits, humans, and animals. This conceptualized structural universe is important in Dumont's understanding of the Indian caste system (Dumont, 1970:43-4), a system in which the whole governs the parts, and this whole is conceived to be based on an opposition. The Ongees and other related tribes of Andaman Islanders do not exhibit the intricate social stratification evident in South Asian ethnography. However, Dumont's method of imposing an opposition based on the whole that governs the parts is applicable to Ongee cosmological structure and the cultural system. Dumont's method is particularly useful in understanding certain aspects within Ongee culture such as why, within the hierarchy system, spirits, humans, and animals are subject to changing positions. Within the Ongee hierarchy system, that which is 'encompassed' becomes 'encompassing'. This is what characterizes Ongee cosmology and hierarchy and is the point at which my explanation and limitation of Dumont's construct of hierarchical structure diverges.

The Ongees believe that the death of human beings, loss of weight, and the dispersal of smell are all ways and steps in which humans are transformed into spirits. It is these spirits who then can move anywhere and everywhere. Ongees also believe that although spirits, humans, and animals share a common space, it is through their different capacities of movement that each remains alive within different places within that space. Spirits hunt humans and displace them by killing and taking them away from the island. Human beings, like the spirits, hunt animals and take them away from either the sea or forest. This relationship of hunting, loss of life, and displacement makes it necessary to regard the hierarchical relations within the 'system' not just as 'apparently static' but 'consistently dynamic'.

Leach (1967) suggests that the structures, which the anthropologist describes, are the anthropological conception of structure reflective of a 'real society', which 'is a process in time' (Leach, 1967: 5). The factors that effect movements, smell, and consequent changes in the hierarchical structure are important in this presentation of ethnography. The Ongees' relationship with spirits and animals is not to be viewed as only an 'as if' model (Leach, 1967), but also must take into account a system of relations that actually do exist for the Ongees. For the Ongees the system of relations within the structure changes and is based on the concept of controlling the movement of smell in order to effect the movement of bodies, that in turn brings about changes in relations. Therefore, I intend to go beyond the level of abstraction, typically used by an anthropologist when describing a social structure solely in terms of the principles of organization that unite the component parts of a system and by which structure exists independently of cultural content (cf. Fortes, 1949: 54-60). I intend to present the cultural content and concepts of cosmological structures and hierarchical systems. This conception of structure does not exist only for the anthropologist or in the anthropologist's mind and is not simply logically created and derived or influenced by, say, the African pastoralist or South American native. Rather, it is a structure conceived on the basis of concepts perceived by the Ongees themselves.

Ongees believe that spirits transform human beings into food or fellow spirits; human beings transform spirits into foetuses and transform animals from raw meat to cooked food. These are all concrete cases of changes in the position of individuals within the 'anthropologist's system' as well as the 'system' that is real for the Ongees. Changes of position are brought about through *enakyu?la* ('power'), that makes it possible for elements and individuals within the hierarchical system to be placed, displaced, and replaced within space.

Enakyu?la also effects the interaction and interactional outcome of living as a hunter and gatherer. The Ongees describe *enakyu?la* as the ability to displace or replace an element. For example, a person who pushes a canoe from the beach and into the sea is believed to be using *enakyu?la*. Similarly, the spirit who takes a human being up to the sky and the Ongee hunter who brings his kill back to the camp-site demonstrate *enakyu?la*. These examples of *enakyu?la* involve movement during the course of which an object is transformed. Changes in location also reflect a change in hierarchical position.

Within the Ongee world-view, power is not only related to aspects of change but also to changes in power relations implicit in the hierarchy. Thus, *enakyu?la* pertains both to location and movement attributed to each element that occupies a position in a place within the space. The position of an element and its capacity to move are the attached aspects of power. Each element within this structure wields power in terms of its capacity to move and to replace or displace other elements in the hierarchy.

All Ongee actions involving the exercise of *enakyu?la* are consciously directed towards a particular end. The Ongee's concern with power, and my depiction of that concern, is in no way a power argument concerning 'needs' and 'goals', as in the works of Malinowski (1944) and Parsons (1949,1951). The ethnographic goal here is to present the Ongee's concern with gaining power as a motive in society. An Ongee faces a choice of action, that is, either to have a *talabuka* ('conjunction') to gain power or a *malabuka* ('coincidence') to avoid experiencing power exerted over him. These choices, *talabuka* and *malabuka*, within Ongee culture are articulated and symbolized through *kwayabe* ('smell'). The dispersal and retention of smell defines a series of actual, real, past, present, and future relations.

The term 'symbol' and its use should not be restricted to 'something that stands for something else', or something else 'where there is no necessary or intrinsic relationship between the symbol and that which it symbolizes' (Schneider, 1968). Symbolic analysis within an ethnography has to show how the symbolic aspects of the culture are a reality related to and derived from individuals within the given culture and substantiated by their actions. The symbols and the meaning of those symbols constitute a system because the people in a given culture share these meanings and connect them to one another. How people connect symbols and meanings is important, since symbols are often thought of as 'things', or the Ongee notion of *gobolagnane*, things that make possible returning or bringing things. Symbols may be words or personal, e.g. the *torale*,Ongee spirit-communicator, and *naratakwange*, initiate who visits the spirit world. The concept of symbol, as I use it, also includes actions. Some symbolic acts are more central than others to a particular culture. Such central symbolic acts are typically laden with a greater than usual variety of meanings, that is, they are 'multivocal' or 'polysemic' (Turner, 1967,1969). The central symbolic actions within Ongee culture, *talabuka* and *malabuka*, are not only symbolizing but also actions that, when properly done, achieve the objective of conjunction or coincidence, maintaining thereby the power positions of spirits, humans, and animals, based on the place attributed to them in cosmology and hierarchy.

The symbolic actions of *malabuka* and *talabuka* pertain to smell. Smell itself is also symbolized as 'things', in the form of *gobolagnane*. Ideas related to smell are the basis of the total Ongee conceptualization of their cosmological structure as well as power relations. Within this conceptualization one can trace total information, enabling 'focalization' and 'evocation' (Sperber, 1975: 119) of 'structure of practice' and 'practice of structure' (Sahlins, 1980).

The central core around which power relations are established and re-established is the symbolism of smell. The centrality of smell and its movement in Ongee culture is similar to Stanley Tambiah's (1976) notion of galactic polity, in which the king is like a powerful sun-like object. The king's gravitational pull keeps in orbit, at some distance from itself, an unspecified number of lesser rulers, each a simulation of the leading king. The whole is unified by a field force characterized by both repulsion and attraction. In galactic polity, power, like light, is in the centre of the system, and it loses strength as it reaches the periphery. The symbolism of centrality of power posits that there is a system, even when one cannot locate a specific power 'mechanism'.

The centrality of galactic polity characterizes the description of power and state in South and South-east Asian cultures as having ritual sovereignty. Ritual sovereignty includes symbols and processes that, in the absence of instrumental mechanisms such as taxes and coercion and checks and law within a state, create a domain or a realm of power. A classic example of this is in Clifford Geertz's *Negara* (1980), that focuses on the ceremonial and theatrical creation of the monarch as well as god (especially the aspect of the king's *digvijayan* and movement) (cf. Inden, 1986). For Geertz, the state, a form of power, is constructed by the king and the king is constructed by constructing a god (Geertz, 1980: 124). The king, an incarnation of the holy, is created through ritual. The state draws its power from the symbolic capacity to enchant, though the king remains a symbol of divinity in the cosmological centre. The king in other descriptions of South and South-east Asian societies is an upholder of the social order and fits into cosmology. Spatial cosmoi are often constructed through courts and temples that connect the order of the person, the society, the king, and the world to the gods, as in Hinduism, or to a transcendent state of being, as in Buddhism. These are cosmologies that construe a ruler as mediating between the forces of 'here' and 'there'.

Returning to the Ongee notion of power, its symbolism is central but there is no distinction between the powerful (king and god) and the powerless (subject of the king). The mechanism of power replicates for and is accessible to all. Therefore, there are neither rituals dealing specifically with power nor are there specific individuals who have the power of conducting rituals. Ongee actions pertaining to power go beyond the Durkheimian (1915) distinction of actions into major classes, namely, religious rites that are sacred and technical acts that are profane. Without addressing the controversy between the way Malinowski's placement of magic as sacred (Malinowski, 1948: 67) and Mauss regarding it as profane (Mauss, 1947: 207), my experience with the Ongees makes me question the assumption that the sacred (religious) and the profane (technical) are distinct wholes. Central symbolism and symbolic acts pertaining to smell and power in Ongee society make the religious and technical, sacred and profane, a continuum and not separate categories (cf. Endicott, 1970,1977).

The dynamic aspect between the hunter and the hunted, as embodied in the central symbols of smell and power, escaped Radcliffe-Brown's (1922) account of the Andaman Islanders. Influenced by Harrison (1912,1913), Radcliffe-Brown, in *Andaman Islanders* ([1922]1964), developed the concept of 'ritual value' attached to certain objects that are socially important for secular reasons. From Radcliffe-Brown's perspective, the performance of ritual generates in the actors certain 'sentiments' that are advantageous to the society as a whole.

Within this framework, Radcliffe-Brown interpreted ritual and postulated that human beings always manipulate their thought categories in a consistent way. The meaning of ritual symbols could be discovered by observing the diverse use of that symbol in both ritual and secular contexts. This was perhaps a good methodological idea, but not honest to the ethnographic reality.

As I have said earlier, the central symbolism of smell destroys the anthropological categorization of the ritual and secular. Radcliffe-Brown's failure to comprehend the manipulative system of thought categories consistently leads him to say, in reference to the question he posed to the islanders, about what becomes of man's spirit after death, that all their responses were 'different and inconsistent' (Radcliffe-Brown, 1964: 168). Radcliffe-Brown believes that the Andamanese ideas on the subject to be 'floating and lacking in precision' (ibid.). I regard the 'inconsistency', 'floating', and lack of 'precision' to be that which characterizes the dynamic cosmological and hierarchical structure of the islanders, since there is no one place to which all dead men's spirits go. There is no 'here' and 'there' for humans and spirits. Humans and spirits coexist in a shared space, they have relations of power between them of such a character that humans can become spirits and spirits can become humans. This dynamic was the reason why Radcliffe-Brown's informants give seemingly conflicting answers.

Radcliffe-Brown's informants in 1906 were from different dialect groups. Under the British administration they had been brought to reside at Port Blair in an effort to control the spread of social diseases. Unlike Radcliffe-Brown, my field-work was carried out on one island among only the Ongees. However, the Ongees also gave various responses when I posed Radcliffe-Brown's question about what happens to man after death. The Ongees would however never disagree with what other Ongees had said about where the spirit of the dead person was. Often one Ongee would incorporate another Ongee's idea and further develop it. It then became clear to me that only segments of an idea might be known yet all the segments could be put together. The one constant in the explanations was the sequence of coming in and going out of spirits in different places within the space.

The response of the Andamanese to Radcliffe-Brown's question indicates that for them the *toma* ('spirits' [*tomya* in the Ongee language]) are not located in one position and at one place. Spirits constantly move around, depending on the relative movements of smells, winds, and the Ongees themselves. Since the hierarchy of the hunter and the hunted is subject to shifts and switches, the dynamic nature of the structure in which relations are systematized is subject to change. Since this is the structure that exists for the Ongee, the Ongee expresses it in the same manner. No ethnographer should expect the Andamanese to say, 'This is the earth, there is heaven, and that is hell'. Structure and organization based on hierarchy and relations of power are subject to constant changes. The Andamanese response is not common in the great religions or the religions of the Book. As R. Firth, in his 1955 Frazer lecture, says, 'primitive beliefs about the fate of the soul are usually not polarized, as they are in the great religions . . . most primitive eschatology is dynamic, with plenty of social interaction' (Firth, 1967: 332).

To convey the dynamic nature of structure and the various relations in it, and the ways in which central symbols and actions systematize meanings and concepts for Ongees, I have treated the issue of smell and power in the methodological framework presented by Levi-Strauss in *Totemism* (Levi-Strauss, 1973: 84). I first 'define' the 'phenomena' of space and cosmology 'understudy' as a relation between two or more 'real or supposed' aspects, such as what moves and where it moves, specifically concerning hunter and hunted, smells, and winds. Finally, I consider *malabuka* and *talabuka* as the general object of analysis to derive necessary connections for understanding what *enakyu?la* is in empirical phenomena such as the rituals of the spirit communicator and the initiation of the young men. In the ritual contexts, the multiple combinations of elements and aspects of smell and movement are used to formulate various relations involving spirits and humans in the cosmological structure and hierarchical system.

Cosmology thus represents an aspect of culture in which structure and process meld, revealing cosmosophy. By cosmosophy I mean the philosophical basis and implications of cosmological structure in relation to social structure. Ongee cosmosophy is based on the allocation of power in cosmology—power that is neither evil nor good, but can produce either negative or positive results. Power, once acquired from the spirits, can be used either to counteract or create danger. Cosmosophy thus provides the islander with an orientation by means of which he can either acquire power, or take protective measures against it; both acquisition of power and protection from it affect Ongee movement and their perceptions of the space in which they move.

The time of my field-work went by as the Ongees moved from place to place on the island of Little Andaman. I realized that for the Ongees movements through space, in relation to other things that moved, created the abstract as well as concrete idea of places being different (*see* Pandya, 1990). Instead of saying it is time to move, Ongees would emphasize the place into which they were moving. For Ongees time does not change, but various movements mark a change of places within space. Therefore, the time to move from the coast to the forest is when the coast becomes a bad place and the forest becomes a good place. The forest becomes a good place because it is there that the men and women moving around find food and will not be hunted and captured by the spirits. When men and women move into the forest, the spirits and winds move to the sea and coastal area, making the movement of the Ongees in that area difficult. This kind of explanation is perhaps difficult to translate and transcribe without using our idea of time and the terms pertaining to it.

It is my intention to show how the Ongees relate space to place, spirits to man, nature to the social, with certain underlying principles through which hierarchy, equality, safety, and danger become temporally specified happenings in our perception, but for the Ongees remain movements in place through which coincidence is avoided and conjunctions are brought about between various patterns of movements of the respective elements that map the Ongee cosmology. For the Ongees it is not space and time, or even space-time, it remains only space and movement. The life of the Ongee hunter and gatherer is ordered on the principle of the movement of smell and wind, through which possibilities and probabilities of life are talked about and dealt with.

> We all have smell, so do the animals. We reach the animal and kill it by releasing all of its smell. The winds take the smell away and never does the smell come back—it is death of the animal—success of the hunter—if the hunter lets the wind take away all his smell he stops moving—it is death—spirits take the hunter away!

Ongee life is based on the economy of movements formed through the 'total' (cf. Mauss, 1954) exchange of smells. In this olfactorial and anemological life, the Ongees create situations of inducing and releasing smell, and also conserving and restricting the release of smell. In this lies the continuity of Ongee hunter and gatherer without being hunted and gathered, a balance of life and death between humans and spirits.

In our 'economy of sentiments', emotions and experience, colours remain the prisoner of form, but not sound and smell. Both the sound and the smell of an object always escape—they are active principles. Sound and smell are for us like space, evoking a notion of freedom. Smell is distinguished by formlessness and sound is put in a series to make a form, distinct from noise. We may have a problem with the indefinability and lack of articulation of the formlessness of sound and smell (Gell, 1977). Thus bottles of wine are to be opened to sniff, advertisements of colognes (capable of seduction) are to be scratched, and we have to be concerned about the situation that stinks, aroma that fills, whiffs that flow, as well as techniques to make olfactory functions easy through decongesting medication. For the Ongees, smells are complete within the source and so highly concentrated that it becomes substance like bone. Apart from this, for the Ongees the smell is completed in terms of the capacity to 'fill up' and 'empty out' the context in which the source of smell exists. In other words, all that moves and is capable of moving within the cosmology is subject to smells and winds, both of which move together. The only significant difference, as the Ongees themselves say, is that 'Smell is contained in everybody like tubers are contained in the ground or in the basket, but the winds can never be held, the wind moves anywhere and everywhere'.

The Ongees do not have meanings for smell by distinguishing other smells (as is the case with signs), but they do talk of winds being distinguished, because with the wind's movement is the movement of smells and spirits. Thus the olfactory and non-olfactory context for the Ongees is non-existent, since in all contexts the winds and spirits are present and capable of moving. The spirits and winds for the Ongees acquire the position of sound, the quality of being ever-present becoming almost intense. Just as not having any sound is also a form of music, not having a smell is a way of preventing the spirits and winds from moving close to humans.

In the Ongee world, the experienced cosmology and mapped movements are a discourse on smell and winds, much like the way we talk about sound (formless) and colour (bound by form). For the Ongees, smell is bound to form in each place, and wind is formless, subject to space. Place, where the movement is reduced to a point of security, safety, and nurturance, is opposed to space, which is associated with freedom, danger, and uncertainty (Tuan, 1977). It is our notion of colour that forms the Ongee notion of smell, and it is our notion of sound that forms the Ongee notion of wind, through which the Ongees talk about change in durations, temporal units, and time *per se*. Consequently, all movements, in terms of when they will take place, are subject to smell and wind conditions. The conditions of the winds positions the body with or without smell, and determines the interactional outcome between the Ongees, the animals, and the spirits.

When I searched for an answer to the question of when the groups and families would move (a characteristic of the hunting and gathering societies [Coon, 1971; Service, 1966; Dentan,

1968; Eder, 1987; Howell, 1984]), I realized that movement was not a mere translocationary act within a given space connecting places through acts of movement. In the very act, translocation was the dynamics of winds and smell and an outcome of this was the patterns of social actions and forms of social interactions.

Within Ongee cosmology, human beings and animals have limited potential for movement, and are restricted to movement along the horizontal axis of space to places like the forest and the sea. In relation to this, birds and spirit communicators are different. They, along with animals such as the monitor lizard and the civet cat, are forms of life that can move across land, into water, and also upwards.

These ideas become clear in the Andamanese and Ongee mythology. During my stay on the island, I collected the following Ongee *account* of *tukuree-ye-jujey*:

Myth No. 4

The ancestral spirits became angry when they saw that their children on the island were not eating the appropriate foods in the appropriate seasons and did not even bother to cook those foods. The ancestral spirits punished their children by sending a storm, followed by a flood, to the island. Some ancestors escaped the catastrophe by climbing up a tall tree. During the flood, the monitor lizard [*Varanus salvator*] and his wife, the civet cat [*Paguma larvata tytleri*], managed to save fire by carrying it up a tree. Those ancestors who could not climb the tree became fish and other forms of sea life. Once the water subsided, those ancestors who could not climb down the tree became birds. Only the monitor lizard and the civet cat were able to come down and swim through the remaining water and then walk over the land. Since they could swim and walk, the lizard and the cat were able to keep the fire in a clay pot (*buchu*). They would scare all the other animals away or burn them to death with this fire. Those other ancestors who could walk on land after the flood went to the forest to make their homes.

The threat to fire posed by the flood sets up the primary dialectical relationship on the basis of which space is divided by the mode of movement — swimming, flying, climbing up, climbing down, walking. Life is divided by zoological categories and by the movements that create spatial divisions: (a) those who remained in the trees, birds and insects, can fly, and they belong to the world of 'light' (sky); (b) those who climbed up and down and can both swim and walk, the cat and the lizard, belong to the world of 'shadow' (land); and (c) those who could not climb up or move down, fish, belong to the world of 'darkness' (sea). (*See* Fig. 4.)

The myth distinguishes the lizard and the cat from all other animals because they can move vertically or horizontally. The characteristic movements of the lizard and the cat are directly replicated among humans: the spirit communicator (*torale*) and spirits are different from all other beings since they alone can move on both the vertical and the horizontal axes. Therefore, Ongees say:

> The kekele (civet cat) and the ayuge (monitor lizard), along with the *tomya* (spirits) and *torale*, are identical. They move quietly and rapidly and have the power to move both up and down and across water and land. They are responsible for making all other humans and animals move only *geeroyebe* (on the horizontal axis) and stay in places such as the forest, coast, land, or sea, but never below the water or above the forest.

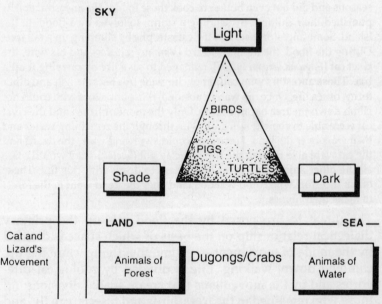

Fig. 4   Categorical Divisions of Life

The Ongee version of *tukuree-ye-jujey* continues by describing how different species of animals, each confined to a specific place and characteristic form of movement, interact because of movement:

Myth. No. 4 (Cont'd)

All the forest and sea animals decided to have a feast at the horizon. The birds and insects were to provide the singing. Since the birds and insects could fly between land and sea, they were to bring various foods. All animals except the monitor lizard and the civet cat were invited. The cat and the lizard found out about the feast because they spotted all the animals painted in red clay going to the celebration. [Red clay paint is applied to the body for all celebrations in the belief that benevolent ancestral spirits will be attracted by the smell.] The cat and the lizard decided to steal red clay paint from the houses of the birds and ants so that they could sneak into the feast. The cat and the lizard ate the red clay and then went into the water for a swim. As a consequence they lost their own body smell and changed colour and, thus disguised, went to the feast. The other animals realized that the lizard and cat had tricked them and were participating in the feast without an invitation. The celebration became a chaotic fight. The lizard and cat started throwing fire at the other animals. The lizard dragged some pigs from the forest to the creek and to the sea and left them there. The cat dragged some turtles from the sea to the forest and left them on tree-tops. The fight became a war. Birds were thrown down from the trees into the sea and made into fish. The crabs did not like this war. So the crabs followed the cat and the lizard everywhere and, by biting the lizard, ended the war. Then the crabs and birds celebrated their prowess in war and the war itself. Again all the animals were invited to a feast to be held on the full moon. The intention of this feast was to reunite the sad pigs [*Sus scrofa andamanesis*], who had been caught in the water when the lizard threw them there with the other pigs from the forest. The turtles, who had been confined to the tree-tops since the cat had thrown them there, would meet with the turtles [*Chelonia mydas* and *Eretmochelys imbricata*] from the water. This was a moment of great reunion and happiness. The pigs and turtles were covered in red clay paint so that their body smell would ooze out and attract their relatives to come and meet them. The pigs and turtles caused much movement between land and water and between forest and sea. Pigs from the sea were rushing towards the forest, and turtles from the forest were rushing towards the coast. The land started sinking down and the water started rising up because of all the animals who were trying to meet their relatives from the land and the water. In all this commotion and excitement only some pigs left the sea to go to the forest and only some turtles succeeded in leaving the forest to live in the sea. Under the light of the full moon, the falling of the land and the rising of the water caused some pigs and turtles to become trapped at the creek and the coast. Those who were trapped between the land and the sea were those who had run to receive their relatives. Pigs from the forest who had gone to receive the pigs from the sea and turtles who had come from the sea to receive the turtles from the land were trapped and became the sufferers of pain, since they had tried to run in the wrong direction. As the night ended, at daybreak, the land stopped rising up because the cat and lizard, who had also been celebrating, ended their dancing. When the cat and lizard ended their dance, the turtles and pigs trapped at the creek became dugongs [*Dugong dugon*]. Since then, at every full moon some dugongs try to leave the creek to become turtles and some dugongs try to leave the creek to become pigs. Their movements cause the water of the creek, the coast, and the sea to rise high and fall low. Some succeed, and those who don't continue to live as dugongs, swimming like turtles but eating like pigs.

In the mythological context space is constituted by movement. The movements of the lizard and cat in relation to the movements of the other animals construct space along the vertical and horizontal axes. The movements of the other animals also distinguish space into different places. The Ongee translocatory seasonal movement between coast and forest sustains the myth's social distinctions of place as being either safe or dangerous. Within this classification of various elements of Ongee cosmology, based on the elements' capacity to move, all humans and animals are in lower positions compared to the hierarchical positions of the spirits, winds, and birds. Spirits and winds are particularly elements of significance, that can move any and everywhere within Ongee space. Within this system of the hierarchy of movement, a certain amount of flexibility and change of position is attributed to each element and is visible through the emission of smells, which are carried to the spirits who are the prime absorbers and receivers of smell. (*See* Fig. 5.)

This movement of smell and wind sets up the possibility of equality within the hierarchical system. The equality is evident in that, though the spirits' movement is greater in its range, by absorbing the smell of humans, the spirits transform humans into spirits. When the spirits absorb the smell contained in the human body, they cause death. In the same way, spirits become humans through the process of birth, in which the condensed

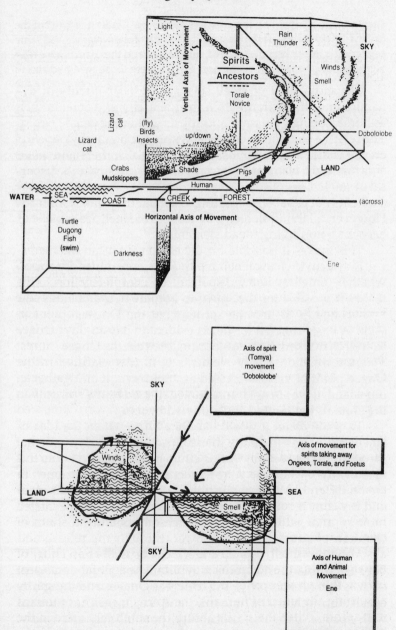

Fig. 5   Ongee Space, Elements, and Forms of Movements

smell form of the spirit is absorbed by women. In the Ongee world-view and in their ontology, spirits and humans, though hierarchically differentiated in terms of their capacity to move, also become equal through the movement of smell.

The movement of smell sets up the basic quality of dynamics and distinction within the hierarchy of elements such as human beings, animals, and spirits within the Ongee world. The relative position of elements within the hierarchy depends on the characteristic opposition of smell absorbers and smell emittors. Each element characterized as living, limits its dispersal of smell. Loss of smell transforms living elements into dead elements not capable of emitting smell but only of absorbing it. However, all living elements by the loss of smell are capable of becoming non-living/dead elements.

This makes it possible for the elements within the hierarchical system to change their relative position, depending upon whether smell is kept, dispersed, or completely lost. This makes it possible for the spirits to become human beings and human beings to become spirits from the Ongees' point of view. A basic concern for this is evident in day-to-day activity as well as marked moments of rituals within the Ongee camps. This interrelationship of elements is further clarified in the Ongee point of view expressed as, 'Spirits hunt and gather in our island; if we are not hunted and gathered by the spirits then the animals get hunted and gathered by us!'

To write about a world like this, I have taken the idea of smell and winds to be the interconnective, underlying theme throughout the sections of this ethnographic account. I intend to show that, while the winds affect and connect with smell to create different situations in the life of the Ongee individual and the community, the basic day-to-day praxis of the Ongee hunters and gatherers is not different from the distinctive duration of ritual. The way power, through the release and absorption of smell, sets up a hierarchy as well as an equality between humans and spirits is replicated on ritual occasions, such as the initiation of young men. The Ongees and the spirits coexist throughout the time span, and the Ongees have to deal with spirits, dangers, probability, possibilities, and safety. They are both hunters and gatherers, and therefore the concerns for movement are common to all contexts. In all situations, the concern to affect smell and the winds form the basis for determining the outcome of the interaction between the spirits and the Ongees. This forms the guiding principle for the structure of this ethnographic account and analysis.

CHAPTER 3

Human Space and Spirit Time: Movements Within Ongee Cosmology

The Seasons and Ongee Space-time

Ongees living on the island of Little Andaman do not consider themselves the only residents of the island. They have always had foreign visitors.[1] In addition to the foreign visitors, the Ongee share their places (*nanchugey*) from time to time with spirits (*tomya*). In the day-to-day activity of hunting and gathering, the individual Ongee regards himself as somebody who is related to the spirits with whom the island is shared. Sharing the island with the spirits is made a reality for the Ongees by the blowing in of the winds from different directions throughout the year, bringing with them various kinds of spirits. The movement of the spirits and the winds make the seasons an experienced reality. Consequently, each season is named after the spirit who comes in with the wind that blows from a particular direction and continues for a particular duration. The spirits, like the Ongees, come to the island to hunt and gather and take the limited food resources of the island, thus making food either available or unavailable. It is because of the food-seeking spirits who visit the island that the Ongees say, 'All food is not to be found in the same season and in all places!'

Food, therefore, is subject to seasonal variations for the Ongee hunters and gatherers. When the spirits are out at sea hunting and collecting food for themselves, the humans are

---

[1] One of the earliest accounts of such a visit is from J.E. Alexander who visited Little Andaman in 1825. The account was published in the *Edinburgh New Philosophical Journal* (1827).

expected to be in the forest, hunting and collecting their own food. The food consumed by the spirits in a given season is proscribed for the Ongees, who cannot gather, hunt, or even consume the same items during that period. Consequently, the Ongee community undertakes seasonal translocations. These translocations are aided by the basic pattern of social organization founded on the division of their tribe into two groups, forest dwellers and coastal dwellers; each identified with its traditional resource base. Constrained by the location and movement of the spirits, each group regularly moves beyond its respective traditional resource base in search of food. During the season, when pig hunting is proscribed, the pig hunters move from the forest to the coastal area and are supported by the turtle hunters. During the seasonal duration when the spirits are in the sea, turtle hunters move from the coast to the forest and are supported by the pig hunters. This pattern of translocation, essential to every seasonal cycle, reflects not only an awareness of spirit movements, which alters the seasons and availability of food, but also signifies the coexistence of the humans and the spirits.

The coexistence of human beings and spirits constitutes the 'nature' from which Ongee derive their basic flavours and sensibilities, which form the cultural constructs of space, space-time, and cosmology. It is within this framework of nature and of cultural constructs that each Ongee is conceived, born, and socialized, and in which he lives and finally dies. The seasons, formed by the movements of the spirits with the winds, are related to aspects of space-time. The spirits' movement from one place to another creates a distinct seasonal duration, which becomes a time unit. However, in relation to the seasonal duration constituted by the spirits' movement from one place to another, humans too move from one place to another. In other words, the determination of the place where the spirits are hunting creates the time when the Ongees move to a place where the spirits are not hunting. Thus, the spirits create time by moving through places and the Ongees also relate to that time by moving through places. The places in which movement is undertaken by men as well as spirits constitute space. It is this mutual relationship which leads to the conceptual and analytical fusion of the categories of time and space into a

unified space-time within Ongee cosmology. Andamanese individuals are connected through the act of movement. Movement generates and defines space and space-time, both of
which effect Andamanese modes of thought and codes of
conduct. Andamanese, and specifically Ongee, ideas about
space and movement seem to parallel those of the Masim
people, whose 'lived world is not only an arena of action, but
is actually constructed by action and the more complex cultural
practices of which any given type of action is a part' (Munn,
1986: 8). In Munn's analysis, the people of Masim give meaning
to the person and a place because of the dynamic actions of
giving and travelling; and thereby the culture connects the
person to place and forms 'space-time' (1986: 9). Munn sees the
subjective dimension, in the forms of remembering and anger,
as a constitutive factor in the formation of 'space-time' and thus
brings out the point that the 'world is experienced by the agents
whose actions produce it . . . producing themselves or aspects
of themselves in the same process' (1986: 11). Munn's analysis
of the range of activities of the Masim is not divided into social
and spatial–temporal, but presents those activities as interrelated 'components of a single symbolic system' (1986: 268).
Similarly, within Andamanese culture *eyolobe* (movements) are
structures that structure space. Bourdieu would regard *eyolobe*
as 'structuring structures' (1977: 90) of the Andamanese cultural construct of space. The notion that the spirits who create
time have an impact on human place coalesces into the notion
of space-time in Ongee cosmology, a world-view in which man
and spirit are interdependent and coexistent. The spirits come
to the island from above or below and create the seasons within
a temporal category.

     In response to the entrance or exit of the spirits from the
island, humans undertake movement from one place to another. This dialectic of movement creates the Ongee map of
space. Spirits live and move along the vertical axis of the Ongee
map. Humans live and move along the horizontal axis. The
vertical and horizontal axes together form the total space in
which both human beings and spirits interact. This interaction
is significant not only because of the food they share but also
because of the created interphases and interstices between the
Ongees and *tomya*. In Ongee cosmology, space-time becomes

the formative foundation for culturally conditioned social interactions. The interaction between man and spirit makes Ongee cosmology a cartographic and kinesic reality that underlies various phenomena in the life and culture of the community.

The Ongees call the closed fist, all five fingers held tightly together, the *monatandunamey*. *Monatandunamey* also generally means the complete seasonal cycle, while literally it means 'my hunger'. The Ongees explain that the five fingers, being connected and interdependent, must go together. 'They are *galawelatetaye* (interdependent), so that the hand can do the work of holding, cutting, and tying'.

However, the Ongees do not count beyond three. For them, any number greater than three is just 'many'. Consequently, the five fingers are not regarded as five but as many fingers. In terms of language use, the thumb and the little finger have distinct terms, but the three fingers between the thumb and little finger are referred to by a single term. The Ongees refer to the thumb by the term *obotabe*, and call the little finger *obeikwele*. The three remaining fingers are referred to together as *oame*. When the fist is opened, the Ongees explain that each finger represents a different season in their experience. The thumb (*obotabe*) is regarded as a representation of Dare, since the pigs become fat and heavy during this season. 'Pigs are just like the thumb—well built!', is the Ongee explanation for identifying the thumb with the Dare season. This idea also makes the season of Dare the prime season for hunting pigs. The little finger (*obeikwele*) stands for the tail end of the Mayakangne? season and the formation of the Torale season, which represents the lean and light pig, which is the least preferred form for hunting. The three remaining fingers (*oame*) are representative of Kwalakangne? and Mayakangne?. The perception of the fingers as together forming a totality explains in part the distinction which the Ongees make regarding the seasons. The Ongee explanation is as follows:

> The three *oame* (fingers) are *gobodegalemba* (season) given to us by *tomya*. *Tomya* give us Mayakangne? and Kwalakangne? . . . *obotabe*, *obeikwele*, and *oame* are all together, like all the *gobodegalemba* together form *monatandunamey* (seasonal cycle) . . . all are *galawelatetaye*? (interdependent). Mayakangne? and Kwalakangne?

[the spirits] come to us along with the winds and spirits. Just when Mayakangne? is to leave the island, before Kwalakangne? can come, Ongees do *gekonetorroka* (ritual) of *getankare*. By doing the ritual we get to *toraleye* (process of extraction) all the honey in a place which is dry, hot, and has no wind no spirits. Ongees get to gather all the honey which spirits get always. This is a *gobodegalemba* which we Ongees make, it is Torale, no winds and no spirits are around and we get all the *tanja* (honey). This is *akwabeybeti* (bad work) . . . spirits are all *beti* (angry and offended): they have no honey to eat. Spirits go and tell Dare. Dare comes to eat the fat pig, but she also leaves since Ongee are then doing ritual of *tanageru* (initiation and killing all the fat pigs). So we again make, Dare *beti* by *gikonetorroka* and bad work of killing all the pigs. Dare leaves the island due to Ongee doing all the ritual and bad work and not sharing food with her. After *tanageru*, when the boy comes back from the home of Dare, Dare agrees to send back Kwalakangne? . . . By doing the ritual of *getankare* we make the spirits and winds leave and stop their coming. Winds and spirits are all angry because it is Torale made by Ongee. We do ritual *tanageru* while the spirit of Dare comes but she too leaves since she gets angry and offended. It is at the end of *tanageru* that Ongee and the *gobodegalemba* which they have made and also start the *gobodegalemba* which are made by the spirits visiting the island.

In light of the Ongee belief that seasons are man-made and also given to man by the spirits, the linguistic and categorical distinction of the *oame* (fingers) as separate from the *obotabe* (the thumb representing the season of Dare), and the *obeikwele* (the little finger representing Torale), becomes understandable. Just as the thumb and little finger are distinct from the remaining three fingers, so the man-made seasons are distinct from the spirit-given seasons. Similarly, the five fingers are interdependent, as are the two types of seasons.

Stopping the winds and offending and sending back the spirits lead to a seasonal change. The Ongees attribute this to the power and efficacy of the ritual *getankare-gikonetorroka*. At the completion of *getankare-gikonetorroka*, Ongees succeed in expelling the spirits and consuming all the honey. With the completion of the *tanageru–gikonetorroka,*Ongees succeed in ushering the spirits and winds back into the shared space. Thereby rituals (*gikonetorroka*) frame the duration of the man-made season within the seasonal cycle. The first ritual (*getankare*) starts the man-made season, marked by Ongees collecting honey and the second ritual (*tanageru*) ends the man-made seasonal duration marked by the presence of winds.

Both the rituals are characterized as *beti*, offensive, and *akwabeybeti*, bad-work. Rituals identical in nature and character start the man-made seasons, i.e. Torale and Dare, and end them too, leading to the start of the spirit-given seasons of Maya-kangne? and Kwalakangne?. Departure of the angry spirits from the island is experienced by the Ongees as the end of strong wind conditions. The end of relatively calm weather, marked by the beginning of strong winds and storms, is indexical of the angry spirits' arrival. A ritual again stops the spirit-given season of Kwalakangne?. Both the ending and starting of spirit-given seasons are framed by rituals, acts of offence, and frame a duration within *monatandunamey*. This framing involves stopping winds, starting the duration of Torale and Dare (man-made seasons). Thus, the duration of man-made seasons is closed on the other end by starting the winds and experiencing the spirit-given season of Kwalakangne?.

The ritual events enable the Ongees to create a period during which the spirits are not present on the island to share the place with them. In a period during which the spirits are absent, the Ongees experience the season of Torale. The absence of the spirits is indexed in Ongee culture by the Ongees gathering all the honey during Torale. Honey is the prime spirit-food, since the jawless spirits particularly enjoy honey.[2] In other words, honey is one of the food items available within the Ongee seasonal cycle which does not need mastication. The

---

[2] The reason honey is considered 'spirit-food' is based upon the major difference between the spirits and the Ongees. The spirits do not have mandibles and hence cannot masticate. This makes it necessary for them to resort to entering fruits and animals to 'suck' their nourishment. Honey is one of the two food items which can be consumed by sucking. The other is cicada grubbies. Both food items, associated with insects, constitute the spirit-food and are available for a short time to the Ongees. The Ongees' special preference for spirit-food is based on the fact that spirit food has a special efficacy, *kama-kuwe-leh-I ekwe* (the capability to produce desired effects). Honey is consumed by spirits, but for the Ongees to consume it they have to undertake special rituals to make the spirits leave the island. This makes the consumption of honey by Ongees very significant. Ongees believe that by consuming honey they become spirits, although this happens only during the Torale season.

ritual of *getankare* sets up the phase of Torale during which the
Ongees can appropriate all the honey and the Ongees' refusal
to share food with the spirits (Myth no. 2, unit J) constitutes the
getankare-gikonetorroka as a bad work and an act of offence.

As the result of *getankare-gikonetorroka*, the man-made seasons continue to constitute the duration of Torale and Dare.
Torale and Dare are ended with the ritual of Tanageru which
involves killing pigs. The killing of pigs is proscribed for the
Ongees while they are collecting honey, and thus makes the
spirits angry. This leads to the return of Mayakangne? and
Kwalakangne? and the start of that phase of the seasonal cycle
when space is shared between men and spirits, and the seasons
are spirit-given.

The idea of offence (*beti*) associated with the rituals is an
element that creates interrelation as well as interdistinction
between the Ongees and the *tomya*. Within the context of ritual
framing, two different types of dynamic relationships between
the Ongees and the *tomya* are created by acts of offence. First,
the ritual stops and limits the movements of the winds and the
spirits. Second, a ritual of the same kind starts the movement
of the winds and the spirits again. This creates a system of
relationships between the Ongees and the *tomya* where one's
act in 'space' has impact on the spirits creating space-time.

The outcome of this system of relationships between men
and spirits leads to the division of *monatandunamey* into two
types of seasons—seasons given by the spirits and seasons
made by men. When the spirits move in from one place to
another within the shared space, they create temporal durations for the Ongees such as Mayakangne? and Kwalakangne?,
which are not just seasons but space-time categories. This is
evident in the Ongee notion that, in the two seasons created by
the spirits, it is important not to encroach and gather where the
spirits are hunting and gathering. Torale and Dare, the seasons
made by men, show that men have an impact on the spirits
through ritual acts, whereby a temporal duration is made
possible when the spirits and the winds are absent from the
place and space in which the Ongees continue to live. Here
again we find that the movement of the spirits and the winds
is the basis of the man-made seasons and space-time categories.

Seasons Experienced: 'Space' Changed by 'Time'

The Ongees move from place to place in relation to the movement of the spirits. The movement of men and spirits together creates various durational divisions within the seasonal cycle. These movements outline and distinguish various spatial divisions such as forest, sea, sky, creek, mangrove, coast, and horizon. The island as space is conceived of as a collection or series of places that are shared by men and spirits. The movement of the winds and spirits, and the movement of the Ongee is interrelated to experiencing seasons, thus making the seasons a spatial experience.

Seasons, as specific to place, are based on where the food sources are found, where the spirits enter the place, and from where the winds blow. Thus the different *gobodegalemba* (seasons), which may be thought of as a temporal duration, are for the Ongees an event experienced in relation to changes in a specific place due to the movements of spirits and winds. The Ongee idea of a seasonal cycle is structured on the series of events happening in different places. What we regard as a time change, is regarded as a spatial change for the Ongee. All the places, *nanchugey* (where events occur, such as hunting pigs in forest, rituals at coast and horizon) together constitute a totality known as *injube*. *Injube* stands for space, a whole of which various places, including the island, are just parts. It is within the space (*injube*) that men and spirits move from place to place (*nanchugey*). The Ongee explanation for translocation is that it is brought about by the interchange of experienced warmth and coolness. As men and spirits move along with the winds and rain, places become different in terms of the experienced temperature. Thus, locations in space are subject to thermodynamic variations due to the kinesics and translocations of winds, spirits, and humans. This, for the Ongees, is an ever-present natural phenomenon referred to as *igagame*. The Ongees explain the natural phenomena of *igagame* in terms of the sun and the moon:

> Sun is wife of Moon. Sun paints her husband's body with red and white clay. She applies more and more white paint on Moon and the husband becomes cold. Thereby the night becomes cold when white-painted husband comes out. In the day the sun remains hot

because she [sun] is making paints for her husband—this is *igagame*.
The change of day to night and night to day, is clay painting
in-between Sun and Moon. This change, *igagame*, is possible from
the Ongee point of view, because winds moving in-between Sun,
the wife of Moon, and Moon, the painted husband, dries up the wet
body paint, leading to an experience of coolness in night and heat
in daytime.

However, the Ongees regard places, space, and *igagame* as
things which, in themselves, do not move. It is the wind which
causes the experience of heat and coolness. The wind passing
through places brings about changes in experienced temperature (including light conditions, such as day and night), and the
Ongees make corresponding changes in location and activity.
The winds are the medium, as it were, which carry the intrinsic
qualities of the sun and the moon (that is, heat and cold,
respectively) from place to place, bringing about changes. All
places (*nanchugey*) comprising the space (*injube*) are characterized to be *elokolake*, an object that is moved by something
else, bringing about change.

Igagame is also regarded as immovable, but it can be moved
by an outside agency. Consequently, change is made possible
by movement, in which an object is moved through different
places by an agent. It is by moving one immovable through
another immovable (that is, *igagame* through *injube*) that the
change experienced in a place is brought about. Ongee explained this distinction in the following manner:

> *Igagame* is like the tubers in a basket and *injube* is like the basket.
> Both the tubers being dug out and the basket to be carried out of
> the forest need an Ongee. What Ongee does in the forest—*tototey*,
> the winds, does in the whole of *monatandunamey* and *injube*. Wind
> carries the *igagame* through the *injube* and the *gobodegalemba* ('sea-
> son') is formed and spirits and man move accordingly.

From the above statement, made by Kunkutaie, one understands that places and space constitute a category of thought
which is not different from the category *igagame*, because all
such categories are here metaphorically referred to as immobile
(*elokolake*).

For Ongee society, places/space change with the passage
of the winds. The change within the Ongee places motivate

their translocationary movements. These movements and their seasonal variations in the hunting and gathering activities offend the spirits and prompt their (the spirits) departures and arrivals. The movement of the spirits establishes the shift of winds creating *igagame*, the force by which change occurs. These movements of both time and space are conceptually inseparable categorical entities. In Ongee cosmology the feature of *igagame*, merging in the features of time and space, establishes change. Both *igagame* and the places forming the space are *elokolake*, immovable; yet are movable by the force of the other, the winds. At any and all times of the seasonal cycle immovable space with places in it is changing because winds carry *igagame* through it. Thereby 'Time' and 'Space' become one in the operative whole, i.e. the Ongee cosmology. Upon this non-distinguishing conceptualization of space and time Ongee cosmology is founded.

Interactional Universe of Ongees and Spirits

The Ongee hunter and gatherer perceives his life and death, his own ancestors, and other spirits, as all existing and moving within the universe/cosmos, shared as the *injube*. In this cosmology, movements order various social relations and interactions between man and spirits. Spirits are distinct from the Ongees in terms of their capacity to move. Spirits alone can move on the vertical and horizontal axes. The movement of the spirits, distinguished from man's movement in an *injube* shared by both, creates the Ongee conception of a person's birth and death. Just as the arrival and departure of spirits on the island has an impact on the Ongee community's seasonal translocation, so also the arrival and departure of the spirits creates birth and death within the Ongee community. Spirits residing up in the sky and below in the sea have to come to the land of the Ongees in search of food. The Ongees, as human beings, cannot go to all places (as living beings), but spirits can. Thereby the whole *injube* has nothing but the potential of travelling spirits and possible encounters with them. The spirits living below in the sea and the spirits living above in the sky do not have all their necessary food resources. The spirits living under the sea have to come to the Ongee land for forest food, which

is primarily pigs, fruits, and tubers. The spirits living below
have only seafood around them. Similarly, the spirits living up
in the sky have only a limited supply of honey and cicadas, but
do not have any supply of forest food or seafood. This makes
both spirits and Ongees hunters and food gatherers. Just as the
Ongee pig hunters and turtle hunters visit each other in accordance with the seasons, the spirits visit the island and go from
place to place to hunt and gather. The spirits who come down
to the island in search of food are frequently trapped inside the
food they consume. When the Ongees consume food in which
spirits are trapped, women become pregnant with the transformed spirits in their wombs. Later, it is the spirit inside the
food and the food inside the woman's womb which causes
utokwobe (birth). As the child grows up within the Ongee community it is seen to be a spirit giving up its spirit identity for a
human identity. This change of identity is completed when the
child gets its first teeth in the lower jaw and starts to walk like
all other Ongee men and women. As the lower teeth emerge,
the child (once a spirit) loses the need to move along with the
spirits and winds since, like other humans, he can now masticate food procured by moving along the horizontal axis. Consequently, a human being is a transformed spirit who has lost
the capacity to move on the 'vertical axis' and has acquired the
capacity to masticate food.

Death (*benchamee*), too, is related to the vertical and the
horizontal axes of movement. In the continuum of life and
death of spirits and human beings, the Ongees say, 'Our teeth
start falling, we grow old and die!' A person's teeth start to fall
if he displeases a spirit. As the person starts losing teeth, the
body becomes light and sick, and finally the relatives, who have
already died and have become spirits, come and take the dying
person away. Since the body is already light, the spirits carry
the dead person away very easily.

However, if a person's death is not witnessed by the camp,
i.e. in the event of an accidental death away from the camp-site,
the Ongees regard it as an improper death. Accidental death,
which no one knows anything about in terms of when it happened and where it happened, is the misfortune of the dead
individual. The person who meets with an accidental death is
regarded to have had an *enegeteebe* (embrace with the spirits);

whereupon the spirits come (on the vertical axis) and take the Ongee away either up in the sky or down below the sea. This event leads to the creation of one more spirit which is not particularly benevolent to the living Ongees. To prevent the creation of 'bad' malevolent spirits, the dead Ongee has to receive a proper burial. Thus, all the dead Ongee become *tomya*, but those who die accidentally and receive no burial service end up as 'bad' malevolent *tomya*.

Just as the spirits who come down to the island may be 'taken in' by the Ongees in the form of food, leading to the birth of a child and an addition to the human community, the inverse can also happen. That is, the Ongees in search of food may be 'taken away' by the spirits, leading to the loss of an individual from the community of Ongees. In such a case, the death of an Ongee is also the birth of a spirit. Consequently, the interaction in the form of mutual transformations of man and spirit, depending upon who is 'taken', leads either to the birth of an Ongee and to the loss of a spirit, or to the loss of an Ongee and to the birth of a spirit. This is one of the metaphysical foundations of the Ongee world-view.

After the natural death of a person, the corpse is buried in the earth. On the night of the full moon following the burial, the son or the son-in-law of the dead person, accompanied by some other men, digs up the burial site. The lower jawbone and other bones are recovered there. This is absolutely essential because by doing so the community of the Ongees ensures that the dead person is transformed into a good and benevolent spirit. The spirits without the lower jawbones are considered to be less hungry and more sympathetic about sharing food with the Ongees. After the lower jawbone has been recovered, the spirits are responsible for coming and taking away the remaining body. In the case of an accidental death, whenever the Ongees are unable to recover the lower jawbone, the spirit created from the dead person becomes a spirit who is always hungry and lacks sympathy towards the Ongees. However, in every death, the movement of the dead body on a vertical axis along with the spirits is bound to take place. Just as the loss of movement on the vertical axis transforms a spirit into a human with a lower jawbone, gaining the capacity to move on the vertical axis by the loss of the capacity to move on the horizontal axis, makes the human into a spirit after death (loss of lower jawbone).

The Ongees reflect their awareness of birth and death in relation to the cosmological construct in various ways. For example, they frequently repeat the phrase, '*Mobetega!*' over again when they have no answer or explanation to offer. The word *mobetega* expresses the idea that 'I have no knowledge or idea, since I am small'. The Ongees correlate the knowledge possessed by an individual with the belief that a person is born once but has to die many times. The multiple deaths of an individual makes him wise, since dying again and again leads to an accumulation of knowledge within the individual. This notion of knowledge, as something that increases with the number of deaths an individual goes through, originates from the notion of *akwanegenegabe* (reincarnation of the dead), especially of the grandparents. Thereby one's dead parents, after dying, become one's children. The term of reference for grandparents and grandchildren is the same. Thus ego would refer to his grandparents and grandchildren as *kolundee*.

In the process of dying and being born, the Ongee has to go through a 'huge tree of *tukwengalako*' that connects all the horizontal places of the humans with the vertical places of the spirits.[3] When birth occurs, the power of the child to be born, in the form of his capacity to breathe and chew, comes down from the tree. At death, the individual's capacity to breathe and chew remains on the island, while the rest of the body goes down the tree with the spirits. The tree marks the prime vertical

[3] The places above in the sky and the places below in the sea are all connected by the *tukwengalako* tree. According to the Ongee, the trunk of the tree passes through the island they reside on. Inside the trunk of the tree is a pathway called *ekwachele*. It is through the *ekwachele* that the powers of masticating and breathing is to be found by the child and lost by the dead person. At the *batitujuney* (horizon) the tree/path bifurcates and every time there is *lololoobe* (earthquake) frequently experienced by the Ongees, they say:

It is earthquake—if there is no bad smell then some dead 'tubers' will rise up and be born as 'fruits'. As soon as the *torale* is over we will have *utokwobe* ('childbirth'). The tuber is going up to come down as fruit—it is on the *ekwachele*—we should have *alankare gigabawe*!

Alankare gigabawe, is a ceremonial singing session held after earthquakes, and is believed to be a means of inviting the spirits to be a part of the Ongee community as children.

axis of movement. The one who has died many times is an individual human who has passed through the tree (the vertical axis of cosmology) many times. By virtue of this repeated movement, that individual gains the capacity to explain and relate human experiences. Consequently, many times in the course of my research, the Ongee would say that, 'the elders cannot answer this question, ask somebody younger who has just come down from the *tukwengalako* tree', implying that the young person has died many times, making him or her truly knowledgeable and endowed with a fresh memory.

Although the patterns of movement are different for humans and spirits, there is an overlap of identity. The 'identity-overlap' is embodied in the Ongee child and the Ongee corpse. A foetus is a spirit (to be transformed into a human being), since it moves on the vertical axis; the corpse is also like the spirit (a human being to be transformed into a spirit), since it moves on the vertical axis with the spirits. To be human and alive means to possess the capacity to move on the horizontal axis.

The spirits and the Ongees thus act upon each other and have an impact on one another's actions. They are related to each other by the processes of birth and death, and are interdependent on each other between birth and death. It is the characteristics and dynamics of act and impact which lends a distinct colour and flavour to Ongee culture and the life of hunters and gatherers in it. Each Ongee person is subject to this act and impact relationship. This makes the Ongee cosmology an interactional universe of humans and spirits.

Spirits and Spirit Places Along the Vertical Axis

Above the land of the Ongees and above the sky, where the branches of the *tukwengalako* tree and the *ekwachele* begin, is the residence of the spirits called Onkoboye?kwa. Spirits like Mayakangne?, Kwalakangne?, and Dare are all neighbours of Onkoboye?kwa. Onkoboye?kwa play a very important role within the Ongee community. At death, all the Ongees become spirits. When the body of the dead Ongee is buried, from the body arises a small human form identical in appearance to the Ongee. This emerging body is called Embekete. Generally, after one or two days the spirits carry away the Embekete to the sky in a spiralling fashion (the movement is called *dobolobolobe*). On reaching the top of *ekwachele* (around the top branches of the *tukwengalako* tree), Onkoboye?kwa start processing the dead Ongee body. This process is called *oyenchemabebe*, which means the safe completion of a journey. The process of *oyenchemabebe* entails placing the Embekete on a hot stone (*ulijojimuera*). After being partially singed, the body is boiled by the Onkoboye?-kwa. The boiled Embekete is placed on a scaffold over rising smoke. Throughout this whole transformation no pain is experienced by the Embekete. The end result of this processing is a change of the Embekete's body into a soft lump of clay. The phase of transformation ends with making sure that no teeth are left on the lower jawbone of the Embekete. Once devoid of lower teeth, what was once an Ongee body and then an Embekete, has become a *tomya*. This also marks the separation of the social individual from the community of the Ongee and incorporation into the community of the Onkoboye?kwa as a spirit.

Onkoboye?kwa marry among themselves and have children born without teeth. They get their marriage partners from the Ongees who are already dead and processed. Onkoboye?kwa are incapable of biting and masticating food, so they come to the island of the Ongee and enter various food substances. This is essential, because there is limited honey and cicada where the Onkobeye?kwa live, and it is not enough to satisfy their hunger. They are to be found in any and every form of food that the Ongees collect or hunt. The Ongees regard it as an insult to eat food containing Onkoboye?kwa, because the latter are actually Ongees *machekwe* ('ancestors'). However, they cannot completely avoid the consumption of limited and seasonal food resources, because the Ongees regard it as important to eat food containing spirits to make women pregnant. This paradox is resolved in the following manner:

> We cannot do wrong to our ancestors who are dead by eating the food which they are eating and are inside it. However we cannot do wrong to our living relatives by not eating anything. So we go to hunt and collect some food, particularly in the given season. This way Onkoboye?kwa have a chance, along with the seasonal spirits, to eat on our island. We do not eat what spirits eat. We eat nothing for a few days when the new season starts and continue to eat food

of the season gone. Spirits look at us—going hungry and then Onkoboye?kwa out of care for us move, eating from one thing to the other. However some spirits remain in the food of the season gone even when the new seasons, winds, and rains have come in.

Above the residential place of the Onkoboye?kwa is the residence of spirits called Goye?go, Gubee-ilemba, and Ekwa-kolodi. All three are responsible for taking away the Ongees and eating them to satisfy their own hunger, after roasting their bodies. Generally, death resulting from accidents provides their food (the dead bodies of the Ongees, not buried by the community, and retaining the lower jawbone). Goye?go, Gubee-ilemba, and Ekwa-kolodi are constantly in conflict with the Onkoboye?kwa. The Onkoboye?kwa, out of love and care for the Ongees try to keep this group of three spirits from taking the bodies of the Ongees who have died in accidents. It is because of this relation between the Onkoboye?kwa and the group of three spirits living above that the Ongees feel that it is important to leave food for the visiting Onkoboye?kwa. Also, the burial practice of recovering the lower jawbone (*ibeedange*) of the dead person and then reburying the corpse is related to the characteristic relations between the Onkoboye?kwa and the group of three malevolent spirits. If the *ibeedange* is retained by the living relatives of the dead person, then the process of transforming the Embekete into the Onkoboye?kwa is expedited. When the relatives fail to recover the *ibeedange*, the malevolent spirits cook and eat the corpse. The dead Ongee then enters the ranks of malevolent spirits and does not become an Onkoboye?kwa. Thus, by retaining the lower jawbones of their dead, the Ongees ensure an increase in the community of benevolent, protective, and caring Onkoboye?kwa. The Ongees regard malevolent spirits as having strong lower jaw-bones, which give them a craving for roasted Ongee flesh. This craving forces the spirits to visit the space occupied by them and to cause accidents intended to increase the population of malevolent spirits. The only way to curtail this is to retain the *ibeedange* of the dead ancestors, thus making an addition to the community of the Onkoboye?kwa, who help prevent accidents.

Between the land of the benevolent Onkoboye?kwa and the land of the malevolent spirits there is the residential place of Eneyagegi and Eneyabegi, the parents of all the living and dead

Ongees. Residing at a place called Inene, Eneyagegi and Eneyabegi keep a watch over everyone who shares the resources of
the forest and sea. Above the land of Goye?go, Gubee-ilemba,
and Ekwa-kolodi is the residential area of the Tetoboah, who
have small teeth and are very skilled in catching fish and
making things that involve the operations of *ulokwobe* (binding
and weaving). They also have the special skills of *ijababe* (cutting). The Tetoboah are responsible for arranging marriages for
the male spirits, and the ones good at binding and tying negotiate marriages for the female spirits. The role of the marriage-
arranger relates the Tetoboah to all the other spirits as *manyube*
(those who arrange and negotiate marriages (generally
mother's brother)). The Tetoboah (like the Ongee mother's
brother) teach everything to the spirits, and in turn are respected by all the spirits. Above the land of the Tetoboah is the
residence of the Jugene who look very different from the Ongees as well as the other spirits. They have facial hair, light-
coloured skin, wear *koylaboi* (clothes), and are skilled only in
binding things and in remaining silent. Above the residence of
all the spirits is the area of Tucenkwaka; these are the spirits
who are always hungry and get nothing to eat. The Tucenkwaka are believed to be Ongees who, while living on the island
among fellow humans, never shared food, so they are punished
by the other spirits and become spirits who die of hunger.

Below the land of the Ongees there is a huge sea in which
the residences of various spirits are arranged in layers. In this
area there is a total absence of sky. Only the sea, with fish and
turtles, is to be found here. Right below the land of the Ongees
reside the spirits known as Eakka. It is in the territory of the
Eakka that the roots of everything stop growing, including
those of the giant *tukwengalako* tree and the path of *tekwachele*
through the tree.

When an Ongee dies of an accident, especially out at sea, it
is the Eakka who transform the dead body. This process involves boiling the corpse, thus causing it to shrink and swell up,
and making it identical with the body of the Eakka. The Ongees
believe that the Eakka are responsible for taking away sick
Ongee. Since the Eakka come from the sea as well as the land
and can acquire the shape of sharks, stingray, moray eels,
snakes, and crocodiles, they are thought to be highly unpre-

dictable in nature. Below the place where the Eakka live is the residence of the Taoere, who are reddish-brown in colour. Occasionally the Taoere, just like the Eakka, come up to the land of the Ongees, but they do not affect the Ongees as other spirits do. The Taoere only come up to the land to get pigs. The bad aspect of the Taoere is that they leave behind a scent in their footprints, which often causes disputes among the Ongees, especially between husbands and wives. The Ongees say that all conjugal disputes are caused by frequenting areas visited by the Taoere and by inhaling the scent of the Taoeres' footprints.

Under the land of the Taoere reside very black, ugly spirits called Tegade? and Toranchu who are afraid of the Ongees and the animals of the forest and sea. Men and animals in turn are afraid of looking at them because they are so ugly and bring bad luck. Tegade? and Toranchu are especially problematic and paradoxical for the Ongee hunters because they scare away the animals. The animals who run away from the Tegade? and Toranchu can sometimes bring *mekwekatakokowebe* ('good-luck') to the hunter, especially if it runs towards the Ongee hunter.

Spirits known as Burage live under the place associated with Tegade? and Toranchu. The Burage have a very fair complexion and extremely long noses, implying a superior ability to smell. Animals of the forest and sea as well as plants regard the Burage as their *mijejeley*, friends, because these spirits, with their acute sense of smell, tell the animals and plants when the Ongees are coming to hunt or gather them. The Burage are therefore regarded as spirits who bring bad luck, since they decrease the availability of food. However, the Burage also trick plants and animals by taking away their young ones, since the young ones are tender to the taste, blaming the abductions on the Ongees. The spirits called Kocheye? reside below the place of the Burage. The Kocheye?, who have large lower jawbones and large faces on their small bodies, cause accidents for the Ongees in or near water. The Kocheye? are also believed to be gluttonous.

All the spirits residing under the land of the Ongees are seen to be residing in the sea known as *kwatanangne*. The winds cannot travel below this sea. Also, the sun, the moon, the stars, and the tides come from this sea.

Within Ongee cosmology, the various spirits reside in all
the places located on the vertical axis, that is, above the land of
the Ongees, up in the sky, and below the land of the Ongees, in
the sea. With the exception of the Ongee land with its distinct
resources (the forest, and the availability of forest food as well
as seafood), these upper and lower places do not have anything
in common. The places above are devoid of sea and forest, and
the places below are devoid of sky and forest. An absence of
the three basic place categories of sky, sea, and forest (land)
makes them distinct from the places on the horizontal axis. It
is only at the place where the Ongees dwell that land, sky, and
sea come together and are ever-present.

Human Places: The Horizontal Axis

Places on the horizontal axis of space are all places identified
with one food resource or another. The availability of food
causes the place to be visited by spirits. Human beings live and
move in the places along the horizontal axis, but spirits only
visit the places on the horizontal axis. It is on the horizontal axis
that the Ongee and *tomya* interact, and experience safety,
danger, birth, and death. The horizontal axis of Ongee cosmology is divided up into three basic sectors: (a) natural division,
(b) residential places for the Ongee, and (c) places where things
of utility and significance are found. In the schematic division
of horizontal space into three sectors, notions about movement,
safety, and danger are reflected.

The outer periphery of the axis is referred to as *ennghame
nanchugey* (natural division), which means literally 'the place
where what is from outside enters in'. The Ongees associate
this sector with the arrival of winds and spirits, a place where
land, sky, and sea meets, a place where outsiders come to, and
a place where confrontations and tensions develop. Residential
areas within the natural division are referred to as *kateta-
belakwe nanchugey*, meaning 'the place where one stays', a place
of rest, a place that is neither hot nor cold, neither safe nor
dangerous. The area of resources forming the core of the island
is called *gakwante-teneyebe nanchugey*, meaning 'the place from
where departure leading to death happens'. It is here that one
meets with spirits. Under each category the Ongees include
various places:

A. *Natural Division*

angage	coastal area
toagege?	creek
ekuju?	swamps
iyele	thick forest
gejegalange	elevated or hilly area
eenge	fresh water source
ingele	sea
kwatule	mangrove forest

B. *Residential Places*

tontebe	sandy stretch in front of the residence (coastal area)
butu	small shrubs and trees in front of the residence (between coastal area and the thick forest strata)
tambojokeo	residential area with cane and dense forest around
totijalo	residential area near the mangrove forest

Just like the vertical axis of Ongee space, the horizontal axis of cosmology has various layers in which reside different spirits. In the same way, on the horizontal axis there are different divisions where different types of resources are to be found. The spirits do visit these resource areas. The following is the constitution of the divisions of Ongee space according to different resources:

C. *Places of Resources (where things of utility and significance are to be found)*

tetoneyekala	place where tubers are found
tanjakala	place where honey is found
tomookala	place where cicada is found
obeedegaleyeh	place where pigs are found
antotene?	place where turtle is found
ambooralugeye?	place where dugong is found
toboreyato	place where crabs are found
nakorolebekala	place where clay for body painting is found
cangaabeh?	place where fire is kept
dabotaabeh?kala	place where iron (scrap) is to be found
ototabeh	place where stones for sharpening the iron tools are found
gejebokala	place where resin is found

ekaanyaneema	place where only men are to be found
	(hunting areas and area where canoes are made)
belakuwebe?ala	place where no-one is found
otanebeynemaa	place where nothing is found

On the horizontal axis, areas associated with resources are regarded to be the central core of the island and around it is the area comprising the residential places. The outermost periphery is identified as the areas consisting of natural divisions and the areas associated with residence and resources are encapsulated within a specific natural division.

Relations of Relations Within Ongee Cosmology

The Ongees in the residential area obtain their subsistence from the resource areas, but this is not by any means a straightforward and practical process. The Ongee hunters and gatherers have an extensive knowledge about what is available where and when. This know-how affects the activity of the entire tribe, but all the activities of hunting and gathering are dependent upon what is happening in the island's outermost area (i.e. the natural division), in terms of the arrival of the winds and the spirits. Resource areas are the places where the distinction between human and spirit movements is most marked, but when men and spirits move into or from the designated places, they affect each other. Consequently, resource areas become places of departure: places from where spirits depart, where an Ongee can lose life, and where accidental death can occur. They are also places where men meet the spirits. This interaction may be positive in nature, or it may be dangerous. These occurrences take place in the space shared between man and spirit through the principle of smell. *Kwayabe*, smell, articulates the relations and interactions between Ongee and *tomya*. It is this *kwayabe* that moves anywhere and everywhere along with the winds which forms the aspect of relations on the basis of which all other relations are formed within Ongee cosmology. The fact that Ongee and *tomya* share places and food resources is to be traced into the Ongee pattern of translocation in accordance with seasonal change; but in this pattern of translocation *kwayabe* is also shared between man and spirit and is the principle upon which the life of the Ongee hunter and gatherer and the culture's classificatory system, as embodied in cosmology, is based. How and what is shared between Ongee and *tomya*, forming a relation of relations, was put very well by Koyra, one of the accomplished spirit communicators on the island of Little Andaman:

> We have to keep going from place to place because all food is not to be found at the same place, since we live with the spirits . . . the more we let the spirits take, the more we get . . . this means going and coming of Ongees and spirits . . . in going to places smell from the body goes out. Body becomes empty and space becomes filled with smell . . . then spirits come in briskly . . . all Ongees die . . . this is a coincidence . . . danger . . . smell should remain in the body and the space should remain empty and incomplete!

Koyra further explained to me how the space is kept unfilled and the smell is kept within its source of emission, that is, the Ongee's own body.

> For Ongees to live and continue moving, *gukwelonone* has to continue. Men hide, *lonone*, their smell. The spirits, *gukwe*, seek, the Ongee. If the *gukwelonone* is good, then very little smell is left behind by Ongee. When spirits come the Ongee have moved away. For this *gobolagnane* are good, they slow down and limit moving spirit and make Ongee moving safe . . . *gobolagnane* makes possible *gukwelonone* between Ongee and spirits. All this is a trick (*togai?kwanenge*). We move and leave little or no smell so that the spirits do not hunt us!

Evident in Koyra's statement is that all places have an element of danger that is inseparable from the life of the Ongee hunter and gatherer. It is a life that depends much on movement and moving around. Moving in relation to shared space makes hunting a process of maintaining a delicate balance between life and death, between man and spirits, and between the hunter and the hunted. In order to kill the animal, which is hunted with bow and arrow, it is equally significant for the hunter to avoid being hunted by spirits. Just as the hunter locates his prey by sight, the spirits can locate the Ongees by tracking down the winds that bring the Ongees' smell. The human body releases the smell that attracts the spirits who can take away the living Ongee. This taking away (referred to as *enegeteebe*) implies a probability that the spirits who are attracted by the Ongee smell will cause the death of a member of Ongee society. The Ongees regard this as something that can happen regularly unless and until countermeasures are taken by the Ongees themselves. The Ongees regard the process by which the spirits track down the source of smell, and take away an Ongee as the process of being hunted by spirits, and refer to it as *malabuka*.

Just as the Ongee hunter carries equipment to hunt with, he also carries *gobolagnane* to prevent himself from being hunted. The *gobolagnane*, which may be clay paints, bones, or fire, basically restrict or limit the dispersal of smell from living bodies. This is an important aspect of hunting. It is the *gobolagnane* which make it possible to hunt animals and keeps the spirits from hunting men by manipulating, restricting, and controlling the smell emitted by the hunter's body.

The hunting of an animal by the Ongee hunter is *talabuka*. This term signifies a deliberate conjunction of the paths of movement between the hunter and the hunted. Since animals too can smell, like the spirits, the animals wish to move away from the approaching Ongee hunter. Consequently, the Ongee hunter has to confine his smell not only from the spirits but also from the animals. However, the occurrence of *talabuka* is regarded as a matter of chance, a possibility, because a *talabuka* takes place only when a *malabuka* does not happen. Avoidance of this coincidence is a way of maintaining life and its balance. Through the avoidance of coincidence those conjunctions become possible that contribute to the sustenance and maintenance of Ongee life. In other words, by controlling the movement of smell, the Ongee hunter succeeds in hunting animals as well as in preventing the spirits from hunting him.

Koyra and his fellow tribesmen go to hunt in accordance with the seasonal proscription and prescription. At sea they use harpoons to hunt dugongs and turtle. In the forest they use bows and arrows along with knives and adze to hunt pigs. For the Ongees hunting is not just a process of getting up and procuring food from one or another place. One never hears the Ongee saying that hunting is a painful, prolonged, and physically taxing activity, which is how I experienced it when the search for animals in the absence of substantial food continued for a day or two. To hunt and gather for the Ongee reflects a special attitude. Though people like Koyra associate the act of hunting with the act of *gukwelonone* (hide and seek), there is also an aspect of power that Koyra refers to as 'trick' (*togai?kwanenge*). This trick involves containing the body smell by means of clay paints and bones, so that animals can be successfully hunted and the spirits fail to hunt the Ongees.

The Ongees are never without *gobolagnane*. For Ongee culture as a whole, the game of hide and seek, involving the trick of releasing and restricting smell, is the basic activity of hunting and unites animals, men, and spirits. It is this very hunting, involving the game of hide and seek and trick, to which Ongees attach a value.

Hunting as an event is not just a life-taking act but also a life-generating activity. This duality of life-generating and life-taking affects the animals, humans, and spirits and relates them to each other. As far as men hunting animals is concerned, the activity involves taking life. In this killing, however, the human community satisfies its basic needs. The outcome of the conjunction of animal and man is the ever-present probability, a happening of coincidence whereby the activity of animal killing may result in the Ongee hunter himself being hunted by the spirits. This is *enegeteebe*, an embrace of man with spirit in the course of hunting, resulting in the Ongee being taken away by the spirits. *Enegeteebe* is a danger with which the Ongee constantly live. Thereby, death in the form of *enegeteebe* as embodied in the explanation of *malabuka*, coincidence, is a regularity. The regularity of coincidence prevails because of the characteristic conception of the body in Ongee culture.

How the conception of body is related to the act of hunting in a very significant way is evident in the terms the Ongees use for hunting. Hunting or the killing of an animal is referred to in the Ongee language as *gitekwatebe*, which means releasing smell resulting in a 'flow of death'. The hunter involved in the act of hunting is referred to as *gayekwabe*, meaning 'one who has his smell tied tightly'. Consequently, hunting in Ongee culture is not just hunting with bows and arrows. It is not just a life-taking and life-generating activity, it is hunting by smell. In releasing the smell of the animal, that is, hunting, the death of the animal becomes possible. This possibility occurs because in the course of releasing the smell from the hunted body the hunter has to restrict the smell released from his own body. Death of the hunted and the life of the hunter becomes possible by means of smell manipulation/management in the process of hunting. *Gobolagnane, malabuka,* and *talabuka* are all indexical of one Ongee idea: smell can be and has to be manipulated. The impact of manipulation of smell is exhibited in the nature and effect of interactions between men, animals, and spirits. The outcome of this interaction determines who hunts and who gets hunted, whose smell is lost and whose smell is retained.

Smell is something that cannot be seen but it moves in and out of *dange,* living things. *Kwayabe* is the Ongees' term for smell and is understood and explained in the following manner.

> Just as the water of the sea comes to land, fills the creeks and then goes back to fill the sea—one never sees who does this filling in and filling back, rising high and falling low our bodies are also filled in and release smell.

Interestingly enough, the Ongee term for the tides is *kwayaye,* the verb form of 'to smell'. *Dange* designates 'all living things', things which have life, as opposed to things which do not have life. This opposition for the Ongees is elaborated on the principle that all living things are capable of one or another form of movement. They release and absorb smell. All *dange* have some part of them which is liquid, they can be cut or broken, they experience change in weight as well as temperature. On the basis of these characteristics attributed to living bodies, the Ongee classification of bodies takes an interesting form (*see* Figures 6 and 7).

Figure 6 represents the way in which things are classified in the Ongee world-view. Figure 7 shows where the bodies are located within Ongee cosmology. It is important to note that within the division of things into various places there is the basic factor that various bodies and their potential and actual smells, and movement within and beyond a place, fixes the bodies in a particular location.

Living things which have a form are to be found scattered in *lime,* and include men, plants, and animals. When the smell of living things goes beyond *lime* into *lichune* (in-between) or *luve* (beyond), the form of living things starts becoming *dange-ka.* The transformation of *dange* into *dange-ka* is brought about

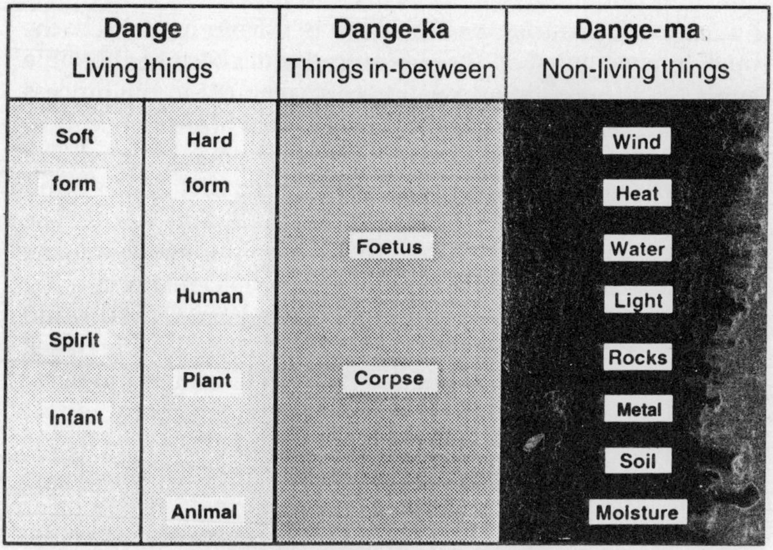

Dange Living things		Dange-ka Things in-between	Dange-ma Non-living things
Soft form	Hard form		Wind
		Foetus	Heat
			Water
	Human		Light
Spirit			Rocks
	Plant	Corpse	Metal
Infant			Soil
	Animal		Moisture

Fig. 6   Classification of *Dange*

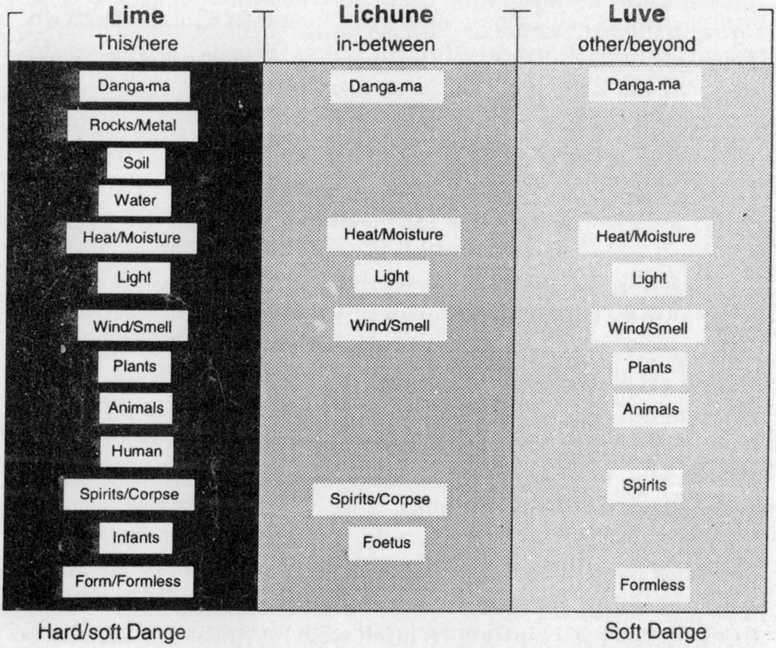

Lime This/here	Lichune in-between	Luve other/beyond
Danga-ma	Danga-ma	Danga-ma
Rocks/Metal		
Soil		
Water		
Heat/Moisture	Heat/Moisture	Heat/Moisture
Light	Light	Light
Wind/Smell	Wind/Smell	Wind/Smell
Plants		Plants
Animals		Animals
Human		
Spirits/Corpse	Spirits/Corpse	Spirits
Infants	Foetus	
Form/Formless		Formless

Hard/soft Dange   Soft Dange

Fig. 7   Body Location in Ongee Cosmology

by *dange-ma*. *Dange-ma* are the agents of transformation, but the most important agent of transformation is a formless *dange*, a spirit. Spirits within this scheme of classification of living and non-living are also capable of moving in all places.

The Living Body and Death

The basic way in which living and non-living forms are conceived and constructed in the Ongee world-view raises questions about the how and why forms change. Smells and winds are compatible with each other because, within the Ongee world-view, both can be felt and have an effect but can never be seen. The Ongees believe that all *dange* constantly release smell and that the winds, which can go anywhere and everywhere, pick-up this smell and take it to the spirits. This makes the winds, the spirits, and smell the main catalysts that bring about all changes ranging from the polar extreme of safety and conjunction on the one hand to danger and coincidence on the other. It is the manipulation of smell which can make a situation safe and life generative or dangerous and destructive of life. All *dange* are believed to have the capacity of releasing smell because of the way they are constituted and conceptualized in Ongee culture.

In the Ongee world-view all living things are perceived as constituted of layers, the central core being hard, which is why bones and wood are to be found in all living things. Around the hard core are the solidified and semi-solidified liquids forming fibrous and flesh-like substance. This flesh and fibrous layer around the hard core makes all living things heavy. The outermost layer of the living body is a covering, that is, a skin, through which liquids are emitted in the form of blood, sweat, and tears. If many liquids escape from the body then a loss of weight is inevitable. This release of liquids from the body is affected by changes of temperature, wind, and moisture conditions. Thus, feeling hot is articulated by the Ongees as, 'I am so hot that I will burn down like ash and become lightweight'. On the other hand, when the body experiences coolness, they say, 'I am shivering, on becoming cold I will fall down like a heavy thing in the sea!' Heat is correlated with lightness and cold with heaviness. Heaviness implies the condensation and hardening of liquids due to a lack of released smell. The sensation of heat is the opposite of this, and induces the release of liquids and smell, thereby making the body light.

Just as the hard internal aspects of the body are related to the solid form of smell within the body, the outermost parts are regarded as being relatively soft, and progress from the semi-liquid to the liquid and unseen form of smell. In other words, the centre of the living body has the hard condensed form of smell. As the body experiences changes in temperature, the solid smell is transformed into liquids and is emitted from the outermost part of the living body, i.e. the skin. The Ongees transpose this distinction of centre and periphery, in relation to movement and thermodynamics, onto movement across the landscape and places. Thus my Ongee friend Totanage, at the end of my field-work, said:

> I will never come with you, I cannot. In your space the winds and
> spirits will find my smell not good to be returned. The further away
> I go from my own place the more difficult it would be for me to be
> cold. The far away places are hot where the winds make smell go,
> my own place is good; it, along with the winds and spirits, keeps
> me well!

If we put Totanage's statement in relation to Figure 7 it seems that the further away one goes from one's own place, the more weight and smell is lost, just as the body, when subjected to heat, loses smell from the centre of the body to the periphery. This is homologous in terms of dangers: the places close to the living Ongee are safe places since there his smell and the winds and spirits are compatible for his personal existence. The further away an Ongee moves from his own place, the warmer it becomes, and danger, in terms of loss of smell, increases. A loss of smell also occurs in places close to the Ongee, but the emitted smell from the body is aided in its return by ancestral winds and spirits. The Ongees explain death in terms of this notion of the 'return of smell'. When an individual dies, it is said that the individual has nothing to breathe since his or her smell did not come back. However, the bones inside the cold corpse are still a source of smell which can be a potential source of smell emission.

Inside the bones in the human body resides another form

of human body. When a person goes to sleep the body form inside the bones goes out. As the external body continues to be motionless in one place, dreaming all the places a person has visited and should visit, the internal body, in the form of dreams, exits and collects all the smell that the external body may have left around in different places in the course of a day. This process, which involves the scattering of smell by the body in the course of movement and the gathering up of that smell by the internal body is called 'spider's home making'. It is for this reason that the Ongees never disturb anybody while sleeping. An interruption in the process of sleeping and dreaming is seen as obstructing the return of the internal body. 'Don't wake him up he is making spider's home, he needs it for hard bones!' was among the first Ongee rules I learned while living with them.

Any pathological disorder, a definite danger, is also explained by the Ongees as a state caused by cuts on the skin and the excessive dispersal of smell, which leads to an acute loss of weight and weakness. The danger of being in this state is associated with the notion that smell attracts spirits and, since the individual's body is not heavy, the spirits are able to pick up and take away the body, thus leading to death and the loss of an individual from the tightly-knit community of the Ongees. It is as a measure against all these possibilities associated with the experience of heat, wind, loss of weight, and dispersal of smell, that the Ongees have evolved their conception of the body. Great importance is attached to keeping bodies cool. When bodies are cool, their smell dispersal is checked and their weight is maintained.

Spirits, *tomya*, are included within the classification scheme as living things like plants, animals, and humans. However, the position of the spirits is unique. Unlike other living things, spirits are formless, that is, they can be felt but not seen. The Ongees say this about the spirits being in close proximity:

> I am shaking due to cold being here . . . someone is embracing me very tightly . . . I feel like a hunted pig . . . but I do not see it.

Spirits, apart from being formless, are also living things that do not emit smells but only receive and absorb them. Unlike other living things, they cannot be cut and do not have bones. Given all these special attributes and characteristics of the spirits, why should they be included in the classified group of living things?

In order to look for answers to this question, we have to look beyond the Ongee concept of the living body and to the interaction of various bodies within the cosmology. What should be done to the body of a dead person is an important matter for the Ongees. This is reflected in the Ongee custom of second burial. In this recovery of the bones, the lower jawbone has special significance. In the Ongee perspective, a proper death requires the appropriate follow-up action. In the case of an individual who is lost in an accident, of which the relatives are not aware, the corpse as well as the mourners and surviving relatives end up in an 'improper' situation. In the case of a person, the time and place of whose death is not known, the bones are not retrievable.

The two forms of death, resulting in the different courses of burial, structure the events that lead to the formation of two distinct types of spirits. Dead Ongees whose bones are retained by the community become 'good spirits', and these spirits, who do not have lower jawbones, are benevolent. The Ongees, whose bones are not retained by their living relatives, become 'bad spirits', or malevolent spirits who have lower jawbones. This distinction of good and bad spirits establishes the basic characteristics of spirits, all of whom were once living Ongees. Both kinds of spirits are particularly interested in consuming food like honey and cicada-grubbies, regarded by the Ongees as spirit food. This spirit food is to be found only in Ongee places.

The good spirits who do not have lower jawbones particularly depend on the Ongees for making provisions of spirit food for them. In return, the good spirits help their living descendants by giving them information pertaining to the forest, the seasons, what to hunt, where to gather, and above all, by protecting the living Ongees from bad spirits. The bad spirits who have lower jawbones are greedy and are ferocious eaters. Their gluttony and selfishness is such that they will even eat Ongees and leave no food like honey for the good spirits. The bad spirits do not want the Ongees and the good spirits to cooperate and survive. The bad spirits who are associated with the act of taking away the Ongees are seen to do so because

they eat Ongee flesh. Since the bones of an individual eaten by
the bad spirits are not with the Ongee society, that individual
becomes yet another addition to the community of bad spirits.

Retaining the bones of the dead relative provides a channel
of communication based on smell between the living and the
dead, connecting various places within the Ongee cosmology.
The retained bones, a form of most condensed smell, are constantly losing a certain amount of their smell, which the ancestors, transformed into spirits, receive. The Ongees control the
dispersal of smell from the bones, and when the ancestral
spirits receive that smell carried by the winds, they come to the
place where the bones are and hence come back to 'their
children'. The retainers of the bones use this efficacy of the
ancestral bones to call upon good spirits, so that the good spirits
can protect the Ongees against the bad spirits who are interested in having an *enegeteebe*, leading to the death of an individual and birth of a bad spirit.

The descent of the good spirits in search of food or for the
protection of the Ongee is also a means by which an addition
in the size of the Ongee community is made possible. On
arriving, the spirits often get trapped in the food found in the
Ongee forest and sea. Later, these spirits trapped inside the
food form the foetuses inside the women that will be born as
children in the Ongee community. For the Ongees there is an
identity between the child and the spirit that is articulated in
terms of the fact that the jawless spirit becomes a jawless child
who too likes soft food and hence depends on its mother's body
for nurturance. Both spirit and child depend on soft and liquid
food and have to satisfy their hunger without masticating food.
This identity of the child and the spirit is further evident in
Ongee culture from the fact that no one calls the child by its
name until it has teeth, and no one tells you the name of the
dead ancestor. The Ongees thus conceive of the human body
as originating from the hunted and gathered food that contains
a formless spirit. The location of the spirit in food and of the
food in a woman leads to the conception of a foetus. Consequently, the idea of hunting is as a life-taking process as well
as a life-generating process.

It is now possible to consider hunting not only at two levels
but at four:

1. The death of an animal by a human hunter.
2. The potential death of a human being caused by a hunting spirit (bad spirit).
3. The continuity of society by means of the procurement of food.
4. The potential impregnation of a woman and the subsequent birth of a child in the community.

One and two form the explicit and implicit aspects of the life-taking aspect of hunting. Three and four form the explicit and implicit aspects of the life-generating aspect of hunting. The bodies of the spirits and the Ongees are thus related. The maintenance of this relation is possible through the manipulation of the movement of smell within the course of all the movements within the Ongee world. Indeed, smell and its movement determine the relation between the Ongees and the *tomya* so that the Ongee hunter hunts without getting hunted.

Food is shared between the Ongees and the *tomya* through the movement of smell and in moving through shared space. Thus, the relation of man and spirit is based on the principle of movement of smell. Through smell the relation of relations are formed, and they lead to the adoption of strategies so that Ongee movements become safe and appropriate. This is particularly important because all living things give off smell, the winds that can move throughout space carry the smell, and spirits are attracted to the places from where smell comes. This binds the Ongees and spirits not only to places in a space but also into a relationship involving rights and obligations. The Ongee myth of *durru* ('thunder') shows how the idea of rights and obligations is related to that of the structuring of relations between men and spirits within shared space.

Myth No. 5

The old winds had not gone very far—we are all very small—everything to eat is available in all the seasons. Ongees eat everything and are not satisfied. They want more and more to eat. All spirits get together and show Ongees that they can eat the hidden food of honey and cicada. [Unit A.]

Ongees like the hidden spirit food very much and they start eating it every day. Spirits tell Ongees when they [the spirits] eat pigs, tubers, and turtle.

Only then can the Ongees eat honey and cicadas, and when spirits eat hidden food Ongees are to hunt and gather . . . All the Ongees started living far from each other and eat without remembering what spirits had told them to eat. [Unit B.]

Spirits got very angry—they all decided to stay on the island making no food available to the Ongee. Hungry Ongees started to cry and sing to Onkoboye?kwa. Spirits got together and decided to send *durru* (thunder, lightning, and downpour) so that we know when we had to live together for safety and heaviness. This way we know when seasons change and give up eating food of the season gone by. [Unit C.]

By staying separately everyone eats everything and soon becomes light and hungry . . . easy for the spirit to carry away such Ongees. Now Ongees stay together—move together from place to place after the winds, season, and spirits have gone through the place. Ongees keep away from the area where spirits are eating so no one is taken away by the spirits. With enough to eat Ongees become heavy and it is difficult for the spirits to hunt the Ongees by smell and take them away. [Unit D.]

Mythological explanations such as that of *durru* form the backdrop to various Ongee actions. For example, when two Ongees meet, this is what they say to each other:

Person A: *Enekutata bangey?* (How are you doing in collectivity?)

Person B: *Kwache na-amborebe.* (Hope you are heavy.)

The term *amboro*, meaning heavy, communicates the idea that one is doing fine. The Ongees believe that to be 'in collectivity', with the band, is safe as well as nourishing because collectivity implies the sharing of food and, as a consequence, the state of being heavy (full of food). Heaviness makes it impossible for the bad spirits to carry away an individual. Implicit in this is the notion that person A is asking person B about his actions and existence in relation to the season and to movement, and about his interaction with the spirits. In case person B wants to communicate that he or she is not feeling well, the response is: '*Geery?-yobe*', meaning, 'I am light'. The root form *gy?ole*, meaning light, implies that one is not getting enough to eat, hence one is not together with the band, and the spirits and seasons are not harmonized with that individual. To be together in a group is regarded as safe. A group is safe not only because of the sharing of food, but also because of the collectivity's release of smell (ref. Myth no. 4, units C and D), whereby the in-

dividuals in the collectivity remain *amboro*, heavy. Smell released by many individuals from one place confuses the bad
spirits who are attracted by it, and all the good spirits are
nearby to protect the Ongees. This prevents accidental death
and forms of bad luck.

The idea of smell is also significant in those actions associated with a meeting. Following the basic greeting and
exchange of pleasantries, the Ongees act out the following
scenario. If a person says that he is heavy, he sits down on the
lap of the person who has asked him how he is, and rubs his or
her nose on the cheek of the inquirer. This ceremonial act of
sitting down in an embrace is called *enegeteebe*, a term also used
for an encounter with the spirits. If the response of the person
is that he is light, the inquirer takes that person's hand and
blows on it. This is known as *gayebabe*. The Ongees describe
the acts of rubbing the nose (known as *nacha-nakobey*) and
blowing on the hand as doing *e?geie kwayabe*, shifting smells,
from one to the other. It is smell, in relation to the winds that
carry it, that binds space and time together, thus relating the
movements of the spirits and the Ongees.

1. Coastal area near Dugong Creek

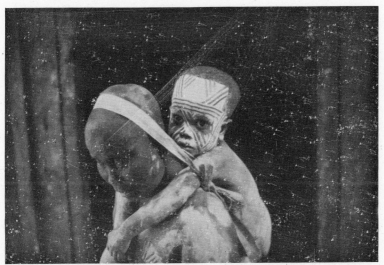

2. Maternal aunt carrying her nephew after his face has been painted
with clay by his mother

3. Typical *korale* set within the forest

4. Returning from a successful pig hunt

5. Initiates on the way to the forest to hunt pigs

6. Novices returning
after hunting pig

7. Initiate sitting on the carcases of pigs during the *tanageru* ritual

8. Shooting arrows into the pig's skull in the course of the *tanageru* ritual

PART TWO

Gobolagnane and *Gikonetorroka*

CHAPTER 4

Models of Movements and the Role of Magical Substances

Danger and Safety

The places that constitute space are sections of the cosmos viewed by the Ongees as *gyambabe*, dangers. All places (constitutive and locative) are arranged in a circular pattern and layered one over the other. Danger exists in every place because of potential encounters with spirits, and such encounters are referred to as *enegeteebe*. The Ongees conceive of *enegeteebe* as being a strong embrace with the spirits, ending in the spirits flying away with an Ongee, like the 'birds fishing on the coast taking off to the sky', emphasizing the diagonal flight-pattern of water birds. *Enegeteebe*, flight, and the carrying away of an Ongee by the spirits, either upwards or downwards, always occurs in a zig-zag pattern and is never straight. In the course of conducting my field-work among the Ongees residing at Dugong Creek, Little Andaman, I encouraged some of the Ongees to use paper and pencils to make 'maps' of the various areas and show the patterns of movements across the regions or places depicted in the map. The idea of having a map is not endemic to the Ongee culture, i.e. the graphical representation of space and route through it. However, the technique of letting the Ongees plot a map in the way 'we' generally perceive a map was a useful 'heuristic tool' to see how the Ongees conceive of the space around them.[1]

In various 'maps' drawn by the Ongees, some of the commentaries accompanying the drawings were particularly important in understanding their ideas about space, places, and

[1] *See* Pandya, 1990 for details.

moving through them. The Ongee word for map is *enechekebe*, which means 'knowing how to move'. Generally, the father and mother's brother orally transmit this knowledge to the younger male hunters and gatherers on expeditions through the forest. The Ongees never draw the *enechekebe* but just describe it. Elaborate terminology is used by the Ongees to describe the patterns of movement from one place to an other. For instance, moving from land to sea is categorically different from moving towards land from sea. However, all the movements that are within the place defined as *kateta-belakuwe-nanchugey* (home-place) are seen to be either a straight path or an angular path of movement. In other words, the movement within the places on the horizontal axis are never perceived to be 'curved' or 'wavey'. On the other hand, the spirits coming down to the places where the Ongees move are referred to as *oatatekoyabe* (which means formation of the *batitujuney*, place where the land, sea, and sky meet, that is, horizon). The Ongees believe that the spirits alone can come down or come up on a vertical axis and then curve onto a horizontal axis of movement, along with the winds. Once they meet with an Ongee or find what they have come to take, the spirits take a diagonal route away from the Ongee's home-place and go up or downwards in a 'wavey' manner which is glossed in the Ongee language as *dobolobolobe*. The term *dobolobolobe* is explained by the Ongee as the 'way in which the snake moves'. *Dobolobolobe* is also the term used for describing the sense of shivering, always associated with being in close proximity to the spirits.

It is evident from the 'maps' made by the Ongees that straight and angular lines for the pattern of movement are associated with places existing on the horizontal axis and have an association with domesticity and humanness. The pattern of movement that is anything but straight is that of the spirits. The movements on the vertical and the horizontal axes of space get connected by means of the distinctive ability of the spirits alone to travel on both the axes and in a *dobolobolobe* manner.

A similar kind of cartographic awareness among the Andaman Islanders is evident in Radcliffe-Brown's work (1964). Radcliffe-Brown referred to the Andamanese magician or the medicine man as having a 'specially endowed power' (*see* Radcliffe-Brown, 1964: 176). How individuals gain this power is illustrated by Radcliffe-Brown in the specific case of Kobo (ibid., 1964: 164), where the medicine man of that name visits the 'world of spirits' and comes back to the forest and his people. Radcliffe-Brown's report of going to the place of spirits making the person distinct, i.e. medicine man / magician is true within the context of the Ongees.

Three practising 'spirit-callers' among the Ongees, known as *torale*, confirmed the recordings of Radcliffe-Brown and reported that when they call upon the spirits, they have to go with the spirits and then come down to the other Ongees. 'All Ongees go down or up with the spirits and come down, but when they do come down they are children—but we the *torale* go up and come down without becoming the foetus in a woman —that is only possible for the spirit and *torale*', said Muroi, one of the practicing *torale* among the Ongees. The Ongees contrast the pattern of spirit departure with spirit arrival. 'Spirits come down to our places like *dankalanka*, helicopters!': that is, spirits arrive by moving on a straight vertical axis of movement and take Ongees away, which is called *enegeteebe*.

Enegeteebe means the death of an Ongee, loss of an individual for the community, and may also result in a situation, an unplanned event, for which no efforts were made. Since the happening is unexpected, it has negative aspects, like misfortune and bad luck, and is referred to as *mekange?getelakwe*.[2] Encounters with the spirits also have a positive aspect, since they generate successful hunting, *makwekatakokokwebe*, the opposite of *makange?getelakwe*. Thus when something positive happens, or something good is found, for which no special effort was made the Ongees see it as an outcome of assistance received from a spirit-encounter.

Makwekatakokokwebe and *makange?getelakwe* form the dual aspect of the occurrence of an *enegeteebe*. Because of this duality the Ongees try to avoid having encounters with spirits, and on certain occasions look up to individuals within the community who specialize in inducing encounters with spirits. Such skilled, influential, and specialized individuals are known as spirit-communicators, *torale*. However, the capacity of doing

[2] Not finding food or losing some article is also regarded as *mekange?getelakwe*.

torale, meaning extraction and involving first a contact with spirits through *enegeteebe*, is limited and exclusive to only a few individuals. Although all Ongees are capable of having an *enegeteebe* with the spirits, these encounters are always negative, as death is inevitable. Only spirit-communicators can have an *enegeteebe* with the spirits and not die. Since the spirits can and do meet the Ongee in every part of the space in which they move, the space around the Ongees is not only shared with the spirits but also becomes dangerous because of possible spirit encounters.

The Ongee hunters and food-gatherers move from place to place in relation to seasonal change. The residential and seasonal translocation makes it possible for the Ongees to hunt and gather in places where spirits are not hunting and gathering at the same time, and the movement of spirits and the Ongees and the changes of season, are all inseparable and inter-affective. So much so that the outcome of an *enegeteebe*, a basic movement event, structures all other movements. Wherever the people decide to camp, through whatever natural region the people move, there is always a potential danger of spirits moving in or moving along with the hunters and food-gatherers. The threat of an *enegeteebe* is a perpetual problematic for the Ongee. It is a situation characterized by the nature of space and the relationship between men and spirits, making every moment of the life of the Ongees open to possible dangers from the spirits. Every movement's beginning and the end of the movement (which completes the act of translocation), are not different from the Ongee point of view. Since there is always a delicate line of separation for the Ongee to be living with his relatives and living with the spirits as a 'dead' person, a delicate and crucial balance must be maintained so that movements do not become dangers and lead to loss. Strategies are evolved by the Ongees to make movements gainful and safe. This involves strategies designed to bring about coincidences or to avoid coincidences of the paths of movements involving men and spirits. The strategies evolved for avoiding *malabuka*, coincidence, and for bringing about *talabuka*, conjunctions, entail the use of *gobolagnane*.

The Notion of Magical Substances

Gobolagnane is the Ongee term for a collection of substances and objects, such as clay paints applied on the body, bones from the body of a dead Ongee, and fire. All the *gobolagnane* are 'given and taken' among all the community members. In day-to-day life, an Ongee may hesitate to ask for food from any or every family (though food is always given away without being asked for), but will never hesitate to demand *gobolagnane*. It is this aspect that makes *gobolagnane* a verb in the Ongee language, meaning 'giving and taking' or 'to go and bring'. Excluding the items under the category of *gobolagnane*, every other thing in the culture is and can be given away but may not be demanded back.

Clay paints, fire, and bones are perceived by the Ongees as having the quality of *kamakulehlekwe* (efficacy), since by possessing these articles the movement of the Ongees becomes a sustained possibility and keeping the spirits away is an experienced reality. The efficacy of the clay paints, bones, and fire is explained by the Ongees as the quality within each object that 'restrains, confines and binds' smell (*kwayabe*) to its source of emission, that is, living things like the human body. Without *gobolagnane* the human body is perceived as being subject to 'cracking up' as a result of the impact of the winds, loss of 'fluids', and the release of smell which attracts the spirits. Thus, the efficacy of the *gobolagnane* lies in its ability to affect the human body and, in consequence, keeps the spirits away.

As Koyra the *torale* describes, movement for the Ongee involves the metaphor of a 'hide and seek' game in which the *gobolagnane* mask the humans from spirits. Indeed, in relation to human smell released from the body that can attract the spirits, the *gobolagnane* function as armour. The Ongees believe that it is crucial to affect the winds because the winds alone go beyond and come from the places where living men and animals cannot go. The Ongees describe the winds as *gwe?yekala*:

We and the animals can go nearby and far-away, and come back.
If we and the animals go beyond far-away, the return is not possible.
It is only the winds which can go and come back from the far-away.
However, in this going and coming of the winds to all and every place, then the smell and spirits too come and go everywhere.

Therefore, winds, spirits and smell are all *gwe?yekala* (like a family who always stay together wherever they go). Winds, spirits, and smell are the only three which can go to the *totoaate?* (place beyond far-away). Since the *tototey*, winds come and go, the spirits, and smell are to be looked after with a great care.

The permeability of the spirits, of smells, and especially of winds within the total space explains the notion of *gwe?yekala*. This notion is even more explicit in a statement made by another Ongee:

> Beyond all the land and the sea nothing can go or come back. It is the end. But beyond this end is the *totoaate?*. One can only see it if one dies or is a good *torale*. I and other men can see it at the time of the *tanageru* (initiation ceremony) but when we come back we forget about it. *Torale* never forgets. The children who come to us from *totoaate?* are also not able to tell us about *totoaate?*. But *totoaate?* is out there because wind comes, wind goes—food is not always to be found in the same place. Spirits come and go—Ongees die and go away. What comes back are young children. All this happens because with the winds, the spirits and the smell go everywhere and anywhere. Winds are *gwe?yekala*.

The Ongee who possesses *gobolagnane* transforms the *gwe?ye-kala* of the winds. This transformation, as a process, is exhibited by those Ongees who always have at least one or the other *gobolagnane* at hand. The *gobolagnane* restrict the smell that is 'cut' and released and carried with the winds, and that communicates to the spirits the location of the Ongee. The presence of the *gobolagnane* makes it possible to avoid *malabuka* (coincidence), since the *gobolagnane* affect the highly permeable movement capacity of smell and spirit in relation to wind which are *gwe?yekala*. With the exception of the *torale*, no Ongee should be without the *gobolagnane*. Even a *torale* avoids the use of *gobolagnane* only in the special situation when he wants the *talabuka* (conjunction) with the spirits to happen. The *torale* explains his technique of creating *talabuka* in the following way:

> When I have to do *enegeteebe* (encounter–embrace) with *tomya* (spirits), I go to the forest without any *gobolagnane*. The winds come—my smell is released—soon the spirits collect around me, they lift me and toss me around. My body cracks up in many places like a bad clay pot. Then the spirits bind me like a pig after it is hunted and carry me away to *totoaate?* and then I have to be on the

watch out for my ancestors at the place where the *tukwengalako* tree ends, the residence of Onkoboye?kwa. It is good for me to have all the pains and thrashing, the more it happens the more smell and body liquids are released, and I can meet many spirits. The more of this happens the chances of my ancestors among the spirits recognizing me are assured and I get help from them to come back.

For the Ongees their conception of space and the potential dangers in it, including the place which is 'beyond and far-away', are very real. It is a profound reality, since no movement is complete without the role of clay paints, fire, and bones, which affect smell and the ethereal quality of winds. This is what makes all movements for the Andamanese a process of transformation leading to avoidance of coincidences or the creation of conjunctions in relation to the spirits. In the use of objects, such as clay paints, fire, and bones, the Ongee hunter and gatherer creates safety as well as danger. He creates danger in terms of having an *enegeteebe* (encounter-embrace) with the spirits and safety by avoiding such encounters.

The significance of the *gobolagnane* lies in its effectiveness and impact on the spirits. This makes even the unseen far-away and places, *totoaate?*, along with the spirits, a reality. In the day-to-day life of the Ongees, the cosmology is not just a 'cultural construct' but a cartographic reality that demands the use of power to manipulate smell, affect the winds, and keep the spirits at a safe distance. No wonder Radcliffe-Brown, in his account, regards things like fire as protective power. It is through the things like fire that the individual, according to Radcliffe-Brown, feels his dependence on the society. According to Radcliffe-Brown (1964: 264), things that are believed to have 'protective power' express a 'social value'.[3] In his view, fire is not the only thing having protective power; ornaments of bones (1964: 257) and clay paints (1964: 255) also have magical/protective powers.

It is evident that in the Ongee belief system, *gobolagnane*, are what Radcliffe-Brown considers as social valuables. It is

[3] 'By the social value of anything I mean the way in which that thing affects or is capable of affecting the social life. Value may be either positive or negative, positive value being possessed by anything that contributes to the well-being of the society, negative value by anything that can adversely affect that well-being' (Radcliffe-Brown, 1964: 264).

true that the *gobolagnane*'s positive powers contribute to the well-being of the society in terms of reducing the possibility of an Ongee dying, and ensures that no 'bad luck' affects the supply of the society's resources and food. The same *gobolagnane* also has the 'negative-value' (ibid., 1964: 264) that places the *torale* in an adverse position when, by refraining from the use of *gobolagnane*, he puts himself in a situation of *talabuka* with the spirits. This *talabuka* can result in the death of the individual *torale* encountering the spirits, but when it is a success, it implies that the *torale* comes back after having an *enegeteebe* and passes on useful information to his community. Generally this information deals with such issues as where to find the resources, where to translocate, what the spirits desire, and what has to be done.

In an analytical consideration of their cultural significance, Radcliffe-Brown's social value and the Ongee *gobolagnane* are the same. However, an expansion of this frame of analysis would be very useful. The *gobolagnane* has social value through its positive protective powers, but it also undermines its own negative value. The negative value of the *gobolagnane* is that by not using it the spirit communicator creates conjunctions with the spirits. The use and disuse of *gobolagnane* are both important. This dual negative and positive value, and the capacity to be manipulated in different kinds of situations, makes the *gobolagnane* powerful. By using *gobolagnane* (the positive protective powers), the Ongees keep the spirits away. By not using *gobolagnane*, the spirit communicator creates a situation in which the spirits are attracted. This makes *gobolagnane* 'socially valuable', because the power within it and the way it is used can either create conjunctions between men and the spirits, or prevent coincidences between men and the spirits. In day-to-day life, the dual aspect of keeping the spirits at a distance and also close to them, or attracting and repelling them, is seen in the various ways Ongee use *gobolagnane* and the ideas they have about it.

Fire as Magical Substance

Never is an Ongee family seen without fire (Radcliffe-Brown, 1964: 472; Man, 1932: 82). Even during a heavy downpour of rain the Ongees are particular about saving burning wood. Fire is important at the camp-site and while in transit; when the Ongees move from place to place they carry fire with them. Having fire is so important that within the forest there are various spots known as *changaabeh?*, which are cleared and a dry log or two is kept lit all the time. The *changaabeh?* are generally located at the points where the boundaries of a re-source division, a residential division, and a natural division overlap.

The Ongees have different names for different aspects and locations of fire. They contrast the heat of the sun, *ikulukutta* (a quality which is implicit in all forms of fire) with *tukuree*, a fire which has smoke and is maintained at a *changaabeh?*. They also make a distinction between two other types of fire having the quality of *ikulukutta*: the fire used in the cooking area, known as *be?yeche*, and the fires used for illumination purposes, called *bonee*. The fine distinctions between fire, heat, and light mark the way in which the Ongees see and create the conception of fire, and explain why they say, 'Fire is truly fire only if it has smoke and in its slow burning the form of *be?yeche* and *bonee* can be made'. Sun (*tonkuloo*) is *ikulukutta* and in it is fire, heat, and light. Linguistically, *tukuree* means 'of *tonkuloo*'. The On-gees regard the sun to be the prime form of fire in the world of spirits. What is distinct about the sun as a form of fire is that 'it has smoke which cannot be seen by Ongee—therefore Ongee cannot be out and close to sun for long durations'. In mythical times[4] this heat and light was brought to the Ongee island, but it did not have smoke. To make fire complete, it has to have smoke, and for this purpose the heat has to be made mild by adding wet wood and resin. *Be?yeche*, cooking fire, is created by adding damp wood and reducing resin. Opposed to *be?yeche* is the *bonee* fire for the purpose of illumination, where no form of wood is added and only resin (extracted by tapping trees and letting the sap ooze out and solidify) is used. *Bonee* does have the aspect of heat associated with it, and is thus different from *tukuree* and *be?yeche*.

It is important for the Ongees to make or replicate the 'sun'

---

[4] The Ongees during the course of field-work narrated myths about bringing fire from this sun that are identical to those recorded by Radcliffe-Brown (1964: 203).

(*tonkuloo*) on their island in the form of *tukuree* (of sun) and to have smoke in it because the smoke of the sun, which is not seen by man, keeps other human beings away. In the same way, by adding wet wood and resin to fire the Ongees create smoke, thus replicating all the qualities of the sun, and this homology keeps the spirits away from humans.

Be?yeche and *tukuree* together are placed in a relation of importance. *Be?yeche*, the cooking fire, is placed within each family's shelter. All the shelters are arranged in a circular pattern with the *tukuree* in the centre. Every time the *be?yeche* is extinguished, a part of a *tukuree* is brought to start it up again. It is therefore the collective responsibility of all the camp-mates to see that the *tukuree* continues to burn constantly. To make *bonees*, resin wrapped in leaves is taken to *tukuree* and ignited. When the Ongees shift from one camp-site to another, or go to the forest for hunting and gathering, a part of the *tukuree* is taken along, just as burning logs or heaps of wet wood form the centre around which the shelters are arranged (*see* Radcliffe-Brown, 1964: 35).[5]

From the camp-grounds, trails radiate out in the directions of the various natural divisions, such as forests, creeks, and seashore. At the point where a trail ends and a natural division starts, the Ongees set up another *tukuree*. Individuals or groups moving on the trails (*echele*) thus have *tukuree* at its two ends. Every time movement is undertaken along a trail, the *tukuree* brought from the camp-site is left at the *tukuree* near the boundary of the natural division, and the one from the boundary of the natural division is taken along while moving through that natural division. On coming out from the natural division, the *tukuree* is left behind at the boundary, and before starting on the trail leading back to the camp-site, another small piece of smouldering wood is picked up. Upon returning to the camp-site, the *tukuree* brought along is placed in the one burning in the middle of the camp-ground. This entire process of taking fire along and bringing it back and of obtaining the cooking fire as well as illuminating fire from the *tukuree* connects all forms of fire to the *tukuree* in the middle of the camp-ground.

These connections continue constantly, and the Ongees are

[5] Cf. E.H. Man (1932: 96).

careful in maintaining the connections and continuity. They say:

> As long as *ikulukutta tonkuloo* is up there it will be day and night and so will remain *tukuree* with us. In *tukuree* being with us all our trails have smoke above and life would continue through moving around.

Ongees believe that the *tukuree* (fire) they now possess was given to them directly by their ancestors. About this fire, the Ongees say:

> The fire we have with us has been with us from the old winds which have passed by (*totekwatta*). Many of the ancestors lived with the power of the fire, whereby safety was assured. Since the fire protected against the possible dangers—it is important to keep that same fire and get more fire from it because it is best for *gukwelonone* inbetween us and *tomya!*

The position of the burning fire at the centre of the camp is the subject of the rule of *gawayabe*. This rule implies that the Ongees should not throw anything into a *tukuree* except damp wood, otherwise the fire (heat) will become so heavy that smoke will not rise from it. The rule of *gawayabe* makes it possible for the Ongees to have a *tukuree* with *gen?yochaye*, smoke. Smoke is important since it forms a layer between the spirits and men moving within space. When the Ongees move from place to place, the rising smoke forms a layer that covers the human smell and forms a screen of smoke that keeps away the spirits who may be attracted by the human smell and who move on a vertical axis within the space. Fire without *gen?yochaye*, smoke, is unable to prevent a spirit–man encounter, and this makes *tukuree* a *gobolagnane*. The fire with smoke not only affects the spirits' movement but also conditions the way in which the Ongees undertake movement.

Walking Behind Fire and Under Smoke

The Ongees always travel in patterned movements when they are outside the camping area. When the men go out to hunt, the women go out to gather, or all the family members go together from one place to another, there is a particular pattern to their movement. The pattern of this movement is predicated on the correct placement of individuals within the ordered frame of single-file order that is always maintained.

Infants who can walk always lead the single-file order of adults. The adults who walk behind the children constantly give instructions about the way to walk, about things to watch out for, and about maintaining a relatively straight line. It is in this situation that children who accompany adults learn about places and about the things to be found in various places. For a child walking in front of an adult, it is the moment to learn about space and places in it and about places and the things in them, which makes him aware of the various movements for the future. The women in the family walk right behind the children. When a family is moving from one place to another, the women are responsible for carrying young infants who cannot walk on their backs. The infants are placed in a loop of bark strip and the loop is suspended from the forehead. Carrying children in this way makes it possible for the women to leave their hands free while undertaking movements over long distances. In their outstretched left hands the women carry burning pieces of wood, *tukuree*. The men walk behind the women. In case no women and children accompany a man or several men, it is essential that the man walking in front of the single file carry the *tukuree*, leaving behind a rising trail of smoke.

Generally, apart from the fire, the person in front is referred to as *melame*, and he or she also carries a machete to cut through the creepers and vines. The *melame* also has a small bag containing some *alame*, red clay paint, and *ibeedange*, lower jawbone ornament. Those who are not related to the family always have to walk between two adults.[6] Whatever the occasion may be, and whoever may constitute the group of people leaving one place and entering another area, everyone has to walk in single file, right behind the person carrying fire. Thus, everyone has to follow the track of the *melame* and hence, walk under a screen of smoke.

The Ongees believe that the *melame* is most knowledgeable about *enechekebe*. In other words, the *melame* is regarded as a

---

[6] As an outsider, working and living with the Ongees, I was always expected to walk in front—constantly getting instructions on how to walk. If I was in a large group I was always expected to walk between two adults.

person who knows his way around different places very well. Since places are well known to the *melame*, it is his responsibility to ensure that the same path is not walked upon during the return journey. In the Ongee language, *melame* as a verb stands for 'making *malabuka*'. Thus, the *melame*, a person, is the guide who conducts people in their movement, and is responsible for creating coincidences, *malabuka*. By walking right behind the *melame*, it is possible to safely avoid the coincidence (a potential of achieving that for which the residential area was left, such as going out to hunt). The Ongees see this making of coincidence and movement not becoming a conjunction as a *talabuka*, because the *melame* walks with fire and people behind walk under the smoke. This is the safest way to move across various places because the smoke rises up to the vertical axis of movement of the spirits, and then goes along with the winds. It is because of this rising of the smoke and of walking under the smoke that safety in transition through space shared with the spirits becomes possible.

The smokescreen left behind by the *melame* creates safety by enabling avoidance of *talabuka*, conjunctions with *tomya*. As a consequence, the *melame* creates a series of *malabuka*, coincidences. Hence the Ongees attribute the success of any mission which entails going out of a residential area and coming to some place without any encounter with the spirits, to the *melame*. Thus, at the end of a successful hunt, the highly preferred part of the animal, the kidney, is given to the *melame*. Along with the regular share of meat, giving the kidney to the *melame* is seen as a token payment to acknowledge that the *melame* alone was responsible for a successful *malabuka*, enabling the hunters to hunt the hunted. In every successful hunt, a *malabuka*, the Ongees also see success in avoiding a *talabuka*, the situation in which the *tomya* hunt the Ongee. Thus, the *melame* is responsible for making food available to the Ongee hunter. In the case of a failure of the *melame*, the Ongee hunter himself becomes the food hunted by the spirit. In cases where the Ongees are just visiting a neighbouring camp, people have to pass through a series of places. The risk of meeting spirits while travelling between the two residential places make the social visit itself a risky undertaking. Consequently, when a movement for a social purpose is accomplished without mishap, the hosts always give a lump of red clay to the *melame*. This act of giving red clay to the *melame* serves to acknowledge the *melame's* capacity to bring about a 'coincidence' between guests and hosts.

In certain ways, the *melame* walking ahead with fire can be regarded as a cartographer. In his ability and success in guiding others safely through the space where spirits move, he connects and disconnects the actual and possible coordinates of various places, so that 'safe' movements are plotted and mapped out for the Ongees behind the *melame*. The *melame* keeps the movement of the Ongees safe and separate from the potential and ever-possible movement of the spirits. The Ongees explain the function and role of the *melame* as follows:

> You need *tukuree* if you are travelling alone— but this is no good. It is always good to move with someone else who is a good *melame* . . . *Melame* knows all about the space and makes possible the *gukwelonone* with spirits. For this *gukwelonone melame* uses his fire and smoke to keep the places of spirits and Ongees disconnected by smell. If this is not done then both the *nanchugey* of man and of *tomya* becomes connected by the *kwayabe*, since the *kwayabe* of the Ongee is always coming out and the *tomya* are always attracted to it. The smoke from the *melame's* fire is like *kugalubete-dane-korale* (spider's web), in which the *kwayabe* of the Ongee body and the nose of the *tomya* remain entangled. *Tomya* looking for us has to look for his nose and we can move from *nanchugey* to *nanchugey* without any danger.

From the above statement it is evident that the *melame* becomes significant because of his capacity to use fire and smoke, by means of which the movement and the places of the spirits and men are kept separate. This separation is brought about by the creation of a 'spider's web'. When the *melame* maps out the path and safety for his followers, he also externalizes his knowledge about space and movement by creating an olfactory–cartographic trap, a 'trick' in which the *tomya* are trapped while the game of 'hide-and-seek' is going on. It is important to walk right behind the person walking in front. This is a rule which is followed not only in the case of children who walk in front without fire, but also in the case of the people walking behind the person carrying fire. This 'stepping over the tracks' left by the individual ahead is seen as a means of leaving and mixing the smell of all the people moving together. This is a measure to ensure that, in case the spirits do trace the smell because the smoke fails to trap the smell of the human bodies and to block the noses of the spirits, it is not just one individual who is traced by the spirits but all those who are together. This strategy thus distributes the risk of attracting spirits by mixing up the smells. It could also be seen as a way in which the Ongees make sure that, in case the smell is not covered by the 'spider's web' created by the smoke from the *melame*'s fire, stepping over the tracks mixes up the smells.

Walking under the smoke and walking on the tracks of the *melame* are both, in principle, means of manipulating and masking human smell. Consequently, Ongee adults are very particular about instructing the children who walk in front without fire, and without the smoke cover, to proceed in a straight line, so that the adults walking behind them may step on their tracks. The act of walking on the tracks left by the person walking ahead is called *gucheyakolabe*. However, when an infant is carried on the back, the adult carries fire and therefore leaves a trail of smoke. The infant on the back of the adult is consequently seen to be under the protection of smoke against spirits. The body of the child is regarded by the Ongee as the most susceptible to being traced by the spirits, since children are believed to be spirits being transformed into the body of an Ongee human. Women's bodies come next to those of children in terms of susceptibility, since women receive the spirits trapped in food, which then change into children in their wombs. The Ongees also see women's menstruation as a means of releasing the smell and signalling the spirits about the Ongees' location. Thus, it is always important to isolate menstruating Ongee women from the camp.

> I have to take my daughter away to *megeyabarrota* (forest) the first time she has *booneeyabe?*, menstrual flow. Otherwise, we all living together will meet the spirits who will come to see my daughter. Once this [menstruation] starts it will be my responsibility to bring material for making her fresh *nakwenage* (grass apron), so that she may have the fresh one after her every *booneeyabe?*. Once my daughter has changed the grass apron, I or her mother has to bury the old one in the *megeyabarrota*. After marriage it will be her husband's responsibility.

The above was the explanation given by a father of a girl who had given a feast for his daughter's first menstruation.

Once again it is evident that not only is Ongee space shared with the spirits, but is also related to by the Ongee as an entity that is spatially extended, contracted, connected, and disconnected in relation to smell. Since smoke has the capacity to break the connections brought about between the space of spirit movement and the space of Ongee movement, every house has smoke in it. The place for the confinement of menstruating women has to have fire, and so does the special shelter constructed for child-delivery, where gaps in the roof thatch ensure that the smoke circulates well over the total camp area.[7] The rules about *gucheyakolabe*, maintaining *tukuree*, and following the single-file order while walking, make fire and smoke 'socially valuable' and *gobolagnane*. Fire and smoke are instrumental in creating safety by restricting the movement of body smell and keeping the spirits from moving into the paths of human movement. As the Ongees point to the smoke rising from the camp-ground they say,

> See the fire is with us and the smoke is along with the winds and the winds are going to where the spirits are, there is a good spider's home made between us and spirits!

To succeed in coexisting with the spirits and in moving through various places for various forms of hunting and gathering, safety is determined by restricting the release of smell from the human body. Keeping fire and smoke close-by embodies the simple logic that they limit the dispersal of smell and restrict the movement of the spirits who are responsible for death and sickness.

Clay Paint as Magical Substance

The *melame* who creates safety for others, by creating coincidence and by personally dealing with any chance encounter with the spirits, is in a most dangerous position. No one walks

---

[7] It is not that the roof of the shelter has gaps so that smoke can circulate because, with the exception of the shelter built for childbirth, the Ongees are comfortable in a smoke-filled *korale* (home) and avoid any attempts at ventilation.

with fire and smoke ahead of the *melame*. Who or what, then, protects the *melame* from spirit encounters?

Although the Ongees who follow the *melame* cover his tracks by means of *gucheyakolabe*, it is insufficient protection. The *melame*, like the other Ongees, ensures his own safety by his (correct) use of other forms of *gobolagnane*, which include clay paints and bone ornaments. Clay paints applied on the body and bone ornaments made from an ancestor's body are also means of restricting the movement of smell and of aiding the movement of the Ongee himself in relation to his space and the spirits in it.

For the Ongees the smokescreen referred to as *kugalubete-dane-korale*, spider's home or a web, is a visible aspect. It is by the 'visible' 'spider's home' that invisible spirits and smell are affected in terms of their movements. Smells and the spirits are never seen, they are only experienced by the Ongees, in terms of feeling cold, light, and dry. Within this frame of reference, clay paints are yet another *gobolagnane* that restrict as well as induce the release of smell. As the fire in the middle of the camp-ground forms the smoke screen above the residential area, preventing dispersal of smell and restricting the movement of spirits, so does the clay paint applied on an individual's body restrict the release of smell. Clay paints (*uwekokey?*) are efficacious since they 'bind' the smell to its source, the living human body.

When a *ganangega-dange* (a body without clay paint) is covered by clay paints, it becomes *gakelekuene-dange* (a body with cuts and marks). Different coloured clay paints and designs on a body are 'cuts' and 'marks' that make the painted body different and distinct from an unpainted one. The range of temperature which is experienced and the way in which smell is released by a painted body differ from an un-painted body. After an Ongee's body has been painted the person declares, 'The clay paint has been good! I feel that my smell is going slowly and in a zig-zag manner like the snake on the ground!'

Alame (red ochre clay paint) and *toleudu* (yellowish-white clay paint)) are the two main colours used for body and face painting. They are used because of their intrinsic qualities. *Alame*, when applied to the body, makes it hot. A body painted with *alame* experiences heat and causes release of *onotangile* (sweat). *Onotangile*, in turn, releases body smell. *Toleudu*, on the other hand, causes the body to cool and consequently minimizes sweating as well as the release of smell. The intrinsic qualities of *toleudu* to cool the human body (*ejemotto*) and of *alame* to heat the human body (*ekullukutta*) are attributed to the location from which the different clay paints are obtained.

All diaphoretic clay paints are associated with ants and the ground on which the ants live. The Ongees explain the intrinsic qualities of different clay paints as follows:

> Ants have home on earth, the soil which we share with them. Generally above the soil one can spot the ants' home. After all the young ants have moved out of their home, the clay from that home becomes *uwekella* (white clay). While the ants make home and the young ones are still inside the *uwe* (clay), the clay is still cold and yellow (*uwekonu-rabanka*). *Uwe* is made into *nu wegerro* (red clay) when the ants go into the home to release more ants. It is the *nu wegerro* which causes a few ants to go into the home but many to come out. Ants get *nu wegerro* from the *tugey* (birds). *Ca?ntembobey* (ants) (*Oecophylla smaragdina*)[8] and *tugey* are good to each other since they help each other. Ants can have more ants because the birds give them red clay, and birds come to know of where the food is from the ants, since birds always keep visiting the spirits in the sky. Ants know where the honey is going to be and the birds learn about it from ants. In due course the birds go and tell the spirits about the honey. To thank the birds the spirits feed them with the red clay and the bird droppings are the way in which the ants get the red clay from birds. This way birds live, ants have children, and spirits know about food in each season.

A regenerative quality is attributed to the red clay. The fact that the availability of honey indexes the hot–dry season, and that it is associated with the supply of red clay, makes *alame* not only heat-generative but also a token of thanks from the spirits to the birds and ants. When an Ongee uses the *alame*, and induces the release of smell from his body, he makes a relationship between himself and the spirits possible. This relationship set up by the *alame* between men and the spirits replicates the relationship set up between the spirits, birds, and ants. Just as

[8] For further details on the prominence of the ants on Andaman Islands, *see* L. Cipriani (1966: 95-8).

smell conveys the presence of the body that emits it to the spirits, so ants and birds give information about food to the spirits. For giving this information, birds and ants get the power of procreation. Spirits give this power in the form of red clay to the birds and ants. Thus, *nu wegerro* starts the transformation of an ant-home (reproduction and multiplication), but the red clay itself also gets transformed into yellow and white clay.[9]

Cutting Bodies in Relation to Smell and Identity

After selecting the soil for the specific colour of clay, the paint is prepared. The large lumps of clay are first crumbled and then saturated with liquids, such as animal fat for red paint and water for white paint. The clay is properly mixed with the liquid and with *oikanare* (spitting saliva) on the palm.

It is always the women who bring the clay from the forest and process it step by step, including spitting into the clay while mixing it for paint. Only women apply body paints. They apply paints to each other and also to the bodies of unmarried relatives. Once a man is married, his wife paints his body. Just before this generous application of paint starts to dry up, a pattern is formed on the skin. The pattern, invariably a permutation of a linear pattern, is created by running three fingers (excluding the thumb and little finger) over the painted surface, thereby removing some of the paint. The pattern is thus formed by paint being removed from some areas, and by being left intact in certain others. The contrast of painted and unpainted skin forms the design. The Ongees not only use the fingers to remove the paint and form the pattern, but also use fine comb-like objects called *jugehy?* (made out of cane strips (*see* Fig 8)). The type of *jugehy?* used depends on the season with which the birth of the body (to be painted) is associated.

The pattern of unpainted and painted skin affects the specific quality of the smell that is released by the body. All human bodies have the capacity to release smell, and each ancestral spirit is identified with a particular type of smell released by a specific living relative. This identification of a spirit and a

---

[9] Cf. Radcliffe-Brown (1964: 192) for association between ants and clay paints.

relative is made on the basis of the amount of smell released. The removal of the paint in the process of forming a design forms a code which the related ancestral spirit, visiting in a specific season, can decode.

The decoding of smell by the spirits and the coding of smell by painting communicates to the spirits the existence of their relatives and brings about a conjunction. The Ongees refer to this communication, which links man and spirit through paint and smell, as a *mineyalange*, literally meaning to remind or to remember. This communication is seen as a positive aspect, and consequently is a *talabuka*, not a *malabuka*. In the process of *mineyalange*, spirits are reminded to protect their living relative against dangers from unrelated spirits.

The Ongees distinguish between the designs painted on the body, *enelukwebe*, and those painted on the face, *enetebe*. The designs painted on the face are always derived from the woman's band identity. Before marriage the design is from the mother's band identity, and after marriage it is from the wife's. The designs painted on the face primarily convey and establish the identity of an individual with the band of association. As a wife prepares to paint her husband's face, she spits into the palm which has all the paint mixed to the right consistency. After spitting into the clay paint, customarily the wife declares:

> Oh! Clay you are mine and about to leave—but you are my own body liquid (*oikanare*, spit). Oh! you clay go—my clay you go to a body that is mine too.[10]

After this 'chant' is completed, the wife paints the face, starting with her husband's forehead. Once the wife has completed the act of spitting and has made the declaration in the form of a chant—about and to the clay—the clay paint is referred to by an exclusive term, *anna-ayube*. In day-to-day

[10] What the wife generally recites is as follows:

Uwe uwe ule-ule! abatta	Clay clay again and again!
meeaba inkebe?	Mine is about to depart
Uwe mea oikanare achibe	Clay of mine body water is good
Uwe uwe ule-ule	Clay clay again and again
Mea Abattal	Mine Oh!
achibe yuvaa	good yours
mea yuga dange!	yours is the body of mine.

usage *anna-ayube* refers to body liquids discharged from the sexual organs and nose of both male and female bodies.

Face painting is based on one of the four designs associated with the four bands of Ongees on the Little Andaman Island. Each woman has the right to use the design that belongs to her matrilineal group throughout her life. Consequently, an Ongee man has his face painted in accordance with the matrilineal design until he is married. After marriage his wife paints the design of her band on her husband's face. This marks the incorporation of the male body into its affinal group and its band identity. The four face designs associated with the four bands are also related to four types of birds. These birds[11] are

[11] Birds have a special place in Ongee world-view. The almost reverential position of birds is based on their ability to communicate with both spirits and men. The Ongee narrate a myth to explain the significance of the birds to young children. The myth goes as follows:

Myth No 6

When the waters rise the birds can climb up, when the winds are strong the birds can go to other places. Because of this, birds can see what all others are doing. Therefore birds and spirits are good friends—birds tell them; and spirits from birds come to know about all. So the birds were in the forest to inform the spirits about what Eneyagegi and Eneyabegi did which led to the creation of Ongees. Spirits then told the birds that they would always say things about others—and since they would do this always, they would never have teeth. Thus birds talked and no one could understand them, but the spirits.

Birds could understand and see a lot but could only tell it to the spirits. So the Ongees were angry. They liked eating the birds. Then the birds decided to become friends with the Ongees. They knew that spirits alone had fire, and if they could bring fire from the spirits to the Ongees then the Ongees would be friendly and not kill them [cf. Radcliffe-Brown, 1966: ch. 5]. Birds decided to fly up to the home of the spirits and steal the fire. Spirits used to keep the fire concealed in shells and throw it down on the forest of Ongees, this was *durru* (thunder and lightning) and the *ayuge* (monitor lizard) used to catch it and keep it from Ongees getting it. This used to happen when the spirits got angry [cf. Radcliffe-Brown, 1966].

Many birds went up. Only *tugegero* [Andaman Scarlet Minivet, *Pericrocotus flammeus andamensis*] succeeded in stealing the fire. In turn, it became black and red—but it did bring the fire to the Ongees. Spirits were quite angry at all this but then *jeetah* [White Breasted Kingfisher, *Halcyon smyrnensis*] taught the Ongee that fish were something that could be had all the year round.

Ongees were all very pleased with the birds. Out of all the birds four—Choulung, Gaye?, Amiya and Amie decided to stay with the Ongees and taught them all about the spirits. We know a lot from the four birds and the people of the four parts where the birds settled started giving and taking

the specific communicators of the welfare of the four bands, in each location, to the spirits. The four painted designs for the face are as follows:

(A) *Choulung*   associated with black-headed bunting
                   (*Emberiza melanocephala Scopoli*)

(B *Amie*   Andaman black-headed oriole
                   (*Oriolus xanthornus reubeni*)

(C) *Gaye?*   associated with great Pied hornbill
                   (*Buceros bicornis homrai*)

(D) *Amiya*   associated with crested bunting
                   (*Melophus lathami*)

The four divisions of the island associated with the four birds are to this day identified with the bands originating, or having *megeyabarrota*.

It is important to note that *Choulung* and *Amiya* both belong to the family of Buntings and are found near the clearings around the coast. They have combinations of red and white plumage. On the other hand, *Gaye?* (hornbill) and *Amie* (Andaman black-headed oriole) are predominantly birds of the deep and thick forest and have a prominent colouring of black and yellow/white, and no red coloration at all. The four birds reflect not only the colour of clay that are accordingly used in face painting but also reflect the basic duality of residing at the coast in the forest.

The body designs, at the same time, are always painted in accordance with the designs associated with the season in which the individual was born. Each individual in the community has one of three painted designs for his body which are based on the season in which the individual was born. The three designs are as follows:

(A) *Barey?* :   design derived from the humpback turtle,
                   for those born in Dare, around mid-May
                   and lasts through mid-July. (This design
                   also is applied to those who are born in the
                   season of Torale, which precedes Dare.)

(B) *Alakaye* :   design associated with dugong, for those
                   born in season Mayakangne?, starting
                   around October and ending by February.

---

from each other—including the clay paints, food, and especially on occasions of *enengelabe* (marriage ceremony).

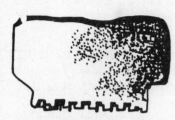

Tagatakwede
Season : Mayakangne?

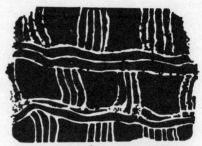

Name of Design: Alakaye

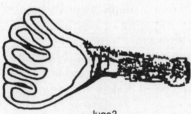

Juge?
Season: Kwalakangne?

Name of Design:
Enetandabo kolale

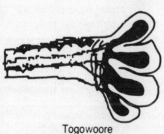

Togowoore
Season: Dare

Name of Design:
Barey?

Fig. 8 Implements Used for Making Clay Paint Designs
(*Enelukwebe*) and Associated Seasons

(C) *Enetandabokatale*:   design associated with the fruit of *daboja*
                                  (*Bruguiera gymnorhiza* [Rhizophoraceae]), for
                                  those born in the season of Kwalakangne?
                                  starting in July and lasting through
                                  October (*see* Fig. 8).

After the age of five, every Ongee has his or her face
painted, and during the season associated with his birth the
body painting is done every day. The painting of the face and
the body is the responsibility of the mother and is also regarded
as a good quality in a wife.[12] Women painting the bodies of
children and husband and of siblings are regarded as important since painting ensures safety and assistance from the
spirits and also marks the individual's membership of a
group.[13]

Curing illness in the Ongee system of medicine deals mainly with affecting body liquids by inducing or restraining the
release of sweat and smell. Body liquids are important because
smell is released from the body through liquids. However, the
curative effects of the different clay paints are regarded by the
Ongees in the following terms: red paint makes the body light

[12] Ongee society and the members of it are divided along the line of gender in
relation to two basic operations. Both the operations *tolakebe* (cutting) and *ulokwobe*
(binding) are integral to the community's existence as a hunting and food-gather-
ing group. Men are responsible for all activities which involve binding, joining,
and tying up. Consequently, Ongee believe that the procurement of food and
cooking is a man's job, since food and eating join the people together in a family.
Women alone are entitled to perform all operations involving cutting, specially
those related to the human body. Thus, cutting hair, cutting the umbilical cord,
and making cuts on the skin to cause bleeding as treatment for pain are all
responsibilities of women. It is because of this basic division that women alone are
responsible for working with the clay paints on the human body.
[13] The face and body painting reflect what Radcliffe-Brown formulated as:

The Andaman Islander ... is made to feel that he is in a world full of unseen
dangers,—dangers from the foods he eats, from the sea, the weather, the
forest and its animals, but above all from the spirits of the dead,—which can
only be avoided by the help of the society and by conformity with social
custom. As men press close to one another in danger, the belief in and fear
of the spirit-world make the Andaman Islander cling more firmly to his
fellows, and make him feel more intensely his own dependence on the
society to which he belongs, just as the fear of danger makes the child feel
its dependence upon its parents. So the belief in the spirit-world serves
directly to increase the cohesion of the society through its action on the mind
of the individual [Radcliffe-Brown, 1964: 327-8].

in weight and white paint makes the body heavy. A change in body weight is the symptom which the Ongees generally report as 'not feeling well'. The successful treatment of illness is referred to as 'cutting by red clay' or 'binding by white clay'. Red paint heats the body and thereby releases smell through the body liquids of sweat. Red paint is believed to 'cut' *ulaateye*, obstacles (a fallen tree on the path, for example, is referred to as *ulaateye*), by releasing liquids and smell. Given this framework of explaining sickness and its treatment, white paint is curative since it binds all the *ulaateye*.

Danger and Its 'Cure' by Clay Paints

Red clay paint and white clay paint are often used for dealing with *ulaateye*, pain. The Ongees distinguish between different kinds and intensities of pain, but they all are seen as being located in a particular part of the body or all over the body. The general notion of pain is that the body experiences weakness and fatigue due to the rapid release of body smell. Since *ulaateye* is also associated with the loss of liquid, decrease of weight, and experience of heat, red paint is not used for its cure. Pathological conditions of *ulaateye* are treated with white clay paint which checks, restricts, and limits loss of weight, liquid, and smell.

White clay paint induces a sensation of coolness. Certain forms of *ulaateye* are thought to be caused by excessive restraint of smell, accumulation of body liquids, swelling, and increase in weight, and sensation of cold. In such cases the individual is treated with the application of red clay paint.

Ulaateye, is diagnosed in Ongee culture through a series of questions and answers. Various symptoms of bodily disorder and discomfort are all classified under one or the other of two typical statements exchanged between the person suffering from pain and the group of people involved in the diagnosis. It is decided only after diagnosis whether the individual should receive white or red paint. Generally the sufferer's closest female relative provides and/or applies the red paint for treatment. The typical statements exchanged are as follows:

Context A

Question　*Konyune-re le be gate? ulatey?*
　　　　　　(What do you think is the meaning/kind of your
　　　　　　'obstacle'?)
Response　*Eahye? meai kuttuakee namborebe Teas-Kwayabe inkenengi
　　　　　　ma ethie lololokobey!*
　　　　　　(Yes! my body is heavy in a big way, body smell is not
　　　　　　leaving at all—so I feel cold!)

Context B

Question　*Konyune-re le be gate? ulatey?*
　　　　　　(What do you think is the meaning/kind of your
　　　　　　'obstacle'?)
Response　*Ahye? meai kuttuakee gerry?-yoba Teaae kwayabe
　　　　　　inkenenegee ma ethie ekullukuttu.*
　　　　　　(Yes my body is light in a big way, body smell is leaving—
　　　　　　so I feel hot.)

People who participated in type (A) context and response
statements would be treated by the regular application of red
paint all over the body. Those who reported statements falling
under context type (B) would invariably have their body
covered with white clay paint.

Ulaateye is not only seen as something within the body
which has to be cured, but is also perceived by the Ongees as
gyambaba (danger). *Ulaateye* is danger, and a person who feels
heavy and cold is bound to die. This death is inevitable because
the winds are unable to carry away any smell and the ancestral
spirits do not come to know that their descendants are ill. It is
important for the Ongees to continue releasing a certain
amount of smell at all times, so that the ancestral spirits are
aware of the well-being or sickness of their living descendants
and help them whenever required. Thus, the release of smell
through the application of curative red clay paint is an invitation to the good spirits to come and help in clearing away the
ulaateye and in restoring order in the body.

When a person's body feels light or hot in a pathological
condition, the state is also dangerous because it can attract bad
spirits. Since *ulaateye* makes a person light, the individual is an
easy target for the spirit's hunt to carry away human beings in
an *enegeteebe*. This is a danger for both the sick person and the

whole community, since the *enegeteebe* leads to the creation of a bad spirit.

In the Ongees' use of different types of clay we find the notions of cutting and binding, which are both strategies for controlling smell. This control of smell is the basis for the cure of the individual, and for the protection of the individual as well as for the collective in situations of danger. The treatment of an individual is also a way of dealing with dangers for the whole society. The distinction between an individual in society and a society for the individual are somewhat fused in Ongee culture. While the individual body states of hot–cold and light–heavy are affected by clay paints, the society paints the individual for its own continuity of existence. This is well demonstrated in the Ongee ritual of initiating young men, where the initiate, the initiators, and the whole community go through various forms of body painting.

Smell, the Channel of Communication, and Paints, the Medium of Communication

Mineyalange, which links men and spirits through paint and smell, communicates the condition of a living body to the spirits. This communication has a dual effect. Both the individual's body-condition and society's state of existence are communicated by smell to the spirits. *Mineyalange* is so significant that clay paints are used specifically to enable the individual body to communicate with all other bodies, including other human beings, animals, and spirits. The significance of clay paints and communication to Ongee society is evidenced in the application of distinct patterns on the body, and the occasion as well as the purpose that are framed by the application of specific paints. After the Ongee have consumed food at any time, especially any form of meat, they must do *nakwanchame*. To do *nakwanchame* means taking some white paint in the hand, mixing it well with water, and applying dabs of it on the mouth, chest, and stomach, and if some paint is still left over, on the hands and legs. The white paint must without fail be smeared on the body after every time meat is eaten. The Ongee explanation for this is:

If I do not do *nakwanchame* the other animals who are yet to be hunted will know that the dead animals were killed [hunted] and are now inside the Ongee stomach, and then they [all the animals] would keep away from the place where we go to look for them and hunt them. Once the animals know that Ongees are the *gayekwabe* (hunters) all that could be *gitekwatebe* (hunted animals) would save themselves. They would hide and trick the Ongee, hunting would become impossible.

From the Ongee point of view, there is a communication (*mineyalange*) between the living animals or unprocured food, the animals yet to be hunted (*gitekwatebe*), and the consumer of the food, the hunter (*gayekwabe*). Communication between the hunter and the hunted occurs because the animal that has already been hunted and eaten, and is present in the form of food in the hunter's stomach, is revealed to the other animals through the air exhaled by the Ongees after the consumption of the meat. After the meat has been eaten, the exhaled breath and the sweat carries the smell of the masticated and swallowed food. The breath, smelling of the eaten meat, mixes with the smell of the body of the person who has eaten the meat. The mixed smell of the hunted and the hunter is thus a threat to the future availability of animals for hunting. It marks danger from the viewpoint of animals, since it indexes their potential for being hunted. The animals become aware that the Ongees are the hunters, and when this happens they 'avoid being in places where the Ongees hunt', thereby causing a failure in hunting. To overcome and prevent this failure, the Ongees control the mixing of the two smells.

It is for this reason that white clay paint is daubed on the skin after a meal of meat. Since consumed meat, which generates heat and sweat, can be limited in its dispersal from the skin, the cooling quality of white clay paint is used. The paint prevents the smell of the hunter's body from mixing with that of the consumed animal. The paint therefore prevents the communication of smell by inhibiting its transmission to the other living animals that are still to be hunted. This enables a deception of the animals to be hunted and also ensures *mekwekatakokowebe*, success in the hunt. This makes *nakwanchame* a medium for bringing the hunted to the hunter through a manipulation of smell, which functions as a channel of com-

munication between the hunter and hunted. The Ongees therefore believe that *talabuka* (conjunction) may not be possible
between the hunted animal and them if they discontinue *nakwanchame*.

The ideas of smell as a channel of communication and the
control of smell by using clay paints are related not only to the
notion of the Ongees as hunters and the animals as prey, but
also to the notion of the spirits as hunters and the Ongees as
prey. The use of clay paints to control smell forms the basis for
the designs which the Ongees paint on the fronts and backs of
their bodies. These designs (*enetebe*) are specifically related to
the season in which an individual is born, and control the
quantity and quality of smell released from the skin. The specific quantity and quality of smell differ according to the painted
design, and are different for each design because each design
leaves a different amount of skin unpainted. The specific form
of the smell formed and determined by the clay paint design is
transmitted to the dead ancestral spirits who coexist with the
living Ongees. This communication is seen as a way to ensure
that the spirits come to their living descendants and protect
them from the bad spirits, who are also attracted by the smell
emitted from the living Ongee body. The painting of the body
of a living Ongee with a specific season's design implies that
the smell released is identical to the smell of the related good
ancestral spirits. The communication of smell to the ancestral
spirits by means of the designs is viewed as different from that
of the *mineyalange* through *nakwanchame* between the hunted
animals and the hunter's body.

Hunting by Smell

Fire and smoke keep the paths of movement of men and spirits
separate and distinct. Smoke contributes to avoiding *talabuka*
(conjunction between the Ongees and the spirits) by confining
the spread of smell, and thereby confusing the approach of the
spirits. Clay paints, like fire and smoke, restrict and aid in the
generation of smell by conditioning the human body. The
painted body designs manipulate smell not only by confining
it but also by generating it. The generation of smell brings about
malabuka (conjunction between man and spirit) and helps in the

occurrence of *talabuka* (coincidence between the Ongee hunter and the hunted animal). *Malabuka* and *talabuka* are the two basic modes of explaining events related to conflicts, confrontation, and loss of life. When the spirits receive the smell released from the moving and living Ongee body, they move so that their axis of movement is coincident with that of the Ongee. The scent can be followed by bad spirits who carry away the living Ongee with them (*enegeteebe*), leading to the death of an individual. Clay paints help to avoid such encounters by ensuring that a particular form of smell is emitted, so that the good spirits related to the living individual also make their paths of movement coincident with that of the living Ongee. To limit the *malabuka* with bad spirits' fire and smoke, forms of *gobolagnane*, are yet another means of protection. Fire and smoke limit the spread of smell from Ongees' living body and check the probable *malabuka*. By using fire, smoke, and clay paints a *malabuka* is successfully avoided, and what becomes possible in this process is a *talabuka*. *Talabuka* generally refers to the conjunction of paths of the Ongee hunter and the hunted animals. For the *talabuka* to come about, the smell emitted from the body of the hunter has to be restricted, so that the animals are unable to trace the possible event of hunting. *Talabuka* and *malabuka* share a common aspect. In both cases, a *malabuka* or probable coincidence between man and spirit has to be avoided, because the Ongee may lose his life. This idea is articulated by the Ongee as, 'If *malabuka* takes place then spirits hunt Ongee!' To ensure that the spirits do not hunt the Ongees, the smell dispersed from the living Ongee body is controlled, manipulated, and limited by means of clay paints and smoke. *Talabuka*, too, has the aspect of loss of life and failure in hunting, but when the successful possible conjunction is achieved, the Ongee hunter successfully hunts the hunted animal. Hunting and loss of life is the common outcome of *malabuka* as well as *talabuka*, but what differs in the two is who gets hunted. In order to hunt animals and not get hunted by the spirits, the Ongees confine and restrict the dispersal of smell from their bodies.

Malabuka and *talabuka* thus demonstrate the delicate balance the Ongees have to maintain between the spirits and the animals, and death and life. Avoiding unwanted coincidences makes desired conjunctions possible, which creates the possibility of living and hunting, while not being hunted. More-over, for the Ongees, the process of hunting and gathering implies sharing all the places and resources in space with the spirits. However, it becomes impossible for the Ongees and the spirits to share space and resources because the smell moves along with the winds. Smells thus are manipulated by humans, while the winds are controlled by the spirits. This makes the humans and the animals in the Ongee world-view different from the spirits, since the spirits can detect smells and the winds are capable of moving anywhere and everywhere.

The way in which smell is controlled, and the purposes for which it is controlled, determine the character of the event of hunting within Ongee culture. All hunting is dependent on smell, and this places humans and the spirits in a relationship of equality, which is the obverse of the notion of hierarchy in the writings of L. Dumont (1980). This equality of men, spirits, and animals is implied in the recognition that altering and manipulating smell determines who hunts what by limiting the dispersal of body smell. That is, in the notions of *malabuka* and *talabuka*, the equivalence of men, animals, and spirits, in terms of the hunter and the hunted is established. *Gobolagnane* like smoke and clay paints are thus the expressive elements which keep the implicitly recognized relations of equivalence distinct and separate. It is the relation of equals within the system of hunting and smell manipulation that makes *gobolagnane* a dominant value, and an ordering principle within the Ongee value system.[14]

Bones as Magical Substances and Classification of Bodies

Just as fire and smoke, along with clay paints, can determine who hunts and who gets hunted, the Ongees also include bones under the category of *gobolagnane*. Bones, too, have to do with coincidences and conjunctions. *Geerange* (bones) are thought to be the prime source of smell emitted by bodies, both human and animal. The bones of dead relatives and the bones of hunted animals are retained and/or prepared. *Gobolagnane*, in

[14] Dumont himself has made the important point that we must not separate value from idea, nor from fact, although his emphasis in making the point is slightly different from the one I am making (*see* Dumont, 1979: 813-14; 1980: 219-23).

the Ongee language, is also a verb meaning 'to go and bring back' or 'come back'. It is this verbal aspect that makes all the paints, fire, and bones identical in value, since they make it possible for the Ongees to move and return without loss of life. Bones in this context actually demonstrate very well how the control of smell makes it possible 'to go and bring back'.

Whether the body is alive or dead, it is the bones that continue to release smell. However, according to the Ongees, the smell emitted by bones cannot be controlled or altered like smell released through skin, which can be affected by the application of clay paints. Once the smell escapes through the skin, its dispersal is limited by smoke. In contrast, the bones release their smell continuously. Bones are the hard condensed form of smell and this is the Ongee explanation for why bones always release smell. The only effects that can be had on bones are to slow down the dispersal of smell by covering the bones and to expedite its dispersal by keeping the bones out in the open.

Geerange (bones) are believed to be the 'hardened' smell of living organisms and cannot be 'cut' but can be 'broken' at the joints. Since bones are strong and solid, various liquids formed by eating food, especially 'spirit-food' tend to cling to the bones and acquire a shape (*tea*). It is this basic principle which the Ongees use to classify all things as *gebokwela* (living) and *enerengewa* (non-living). It is around these two principles that the Ongee organize categories within the classificatory system of living things. All living things are further divided as *dange* and *dange-ma*. All plants, trees, wood, animals, and human bodies are *dange*. All things that are *dange* are capable of being cut open and also of being tied together. However, the Ongees report that insects, reptiles (excluding monitor lizards and crocodiles) are living beings, since they have an aspect of growth, but they are not *dange* since their bodies are associated with eggs that cannot be cut open but simply break. All the 'reptiles' collectively are referred to as *gakhwey?kabe*. It is because of this that the Ongees also regard crabs (*gakuwe*) and *naralanka* (turtle) as being the strongest forms of *dange*, since they have a bone covering (shells) on the outside and retain their smells very well inside the bones. Shellfish (*Nya?-nya*), are also included in this group. Collectively they are called *geinkebe*. The above-mentioned group of living things are subsumed under the category of *dange-ma*. Turtle, shellfish, and crabs, all of which are in a way preferred food and are essentially regarded as food substances that make a person strong, are grouped under the name *dange-ka*. Literally, *dange-ka* means 'where is the *dange*' and *dange-ma* means 'there is no *dange*'. It is the particular significance of the *dange-ka* within the Ongee classification of living things that is highlighted, in that at the end of each ritual and ceremony it is important to eat the meat of things belonging to the *dange-ka* group.

There is a continuity between *gebokwela* (living things) and *enerengewa* (non-living things). *Gebokwela*, by nature, are living things (with bones), which have the capacity to release smell slowly. *Enerengewa* are living things (without bones) that receive and absorb smell. When the covering around the bones, that is the flesh and skin of *gebokwela*, starts decaying and falling apart, the *gebokwela* is said to be in a state of *entengetakabe*. Over a period of time all *entengetakabe* become *enerengewa*. The living Ongees become spirits, *enerengewa*, when they die, but the intermediary stage of *entengetakabe* includes stages and forms of the body like the corpse. This gives us an Ongee classification of living things in which living bodies only emit smell whereas non-living bodies absorb smell but do not emit it. Within the classification of bodies, living and non-living, a spirit is often described as having the following attributes and characteristics:

> *Tomya* spirit is different, separate and not like tree, wood, or humans. Spirit is not *dange*. Ongee, plants, and animals are all *dange*. *Dange* has smell tied to *gebo* (term for flesh of animals and pulp of plants) is formed due to the *genekula* (process of liquids becoming solid). *Geetangey?* (fat) and the *gatee* (skin) all help in keeping the *gebo* to *geerange* (bone) and *kwayabe* (smell) within the *geerange*. When a person dies his smell starts going out, because the *gebo* starts/becomes untied to the *geerange*. Thus after death we only have few bones left, the rest all goes with the winds and *igagame*. *Gebo, geerange,* and *gatee* all fall apart by *nanguchumemy?* (transformed by the decay). When *tomya* had his *genekula* (process of liquids becoming solid) all right he too had *dange*. After *nanguchumemy?* the *dange* becomes *tomya*. *Tomya* can receive the smell from the *geerange* of his *dange* with the living Ongees. So the *tomya* are always coming to the *nanchugey* of the Ongees to get back to the

geerange. If this happens, it is *talabuka* for the *tomya* but sometimes *malabuka* for the Ongees. If everything is *talabuka* then the *tomya* becomes an *ale* (child). *Tomya* acquires all the liquids inside the woman (womb) and his smell all gets *genekula*, and *tomya* becomes child with *geerange* and *dange*.[15]

[Muroi and Tejaie: conversations, December 1983.]

Ibeedange: *Preserved Bones that Cut Smell*

After a person's death his body is referred to as *gegi*, meaning a root or tuber. The body is buried in the section of the forest associated with his or her matrilineal group. In the case of men the body is buried in the forest, and in the case of women the burial takes place near the coast. After burial, on the following full moon, the chief mourner, along with other men, undertakes a trip to the burial site. This is called *jonetalebe-geerange gobolag-nane*. All the participants dig up the ground where the corpse was buried, using digging sticks. Some of the bones are collected carefully and brought back to the camp. The bones recovered from the burial site are sprinkled with red clay paint and kept in a small basket lined with leaves, bark, and sometimes bits of cloth. Bringing back the bones to the camp ends the duration of mourning. The immediate relatives of the deceased person are now permitted to talk and the various food proscriptions become inapplicable.

To mark the end of mourning, a feast is held. It is at this time that the basket with the bones is emptied out. The children, the spouse, and the parents of the dead person take the bones to make *yenagoranka*. Those are ornaments made out of the bones recovered from the burial site, which are kept by the relatives of the dead person. Making the bone ornaments involves tying the yellowish dry stem skin of the *keye?* (*Dendrobium* species) plant onto the bones, in order to cool the bones and to keep their smell within them. Red clay paint is applied

---

[15] Ongees call this transformation of the *tomya* into a child, through the process of solidification of smell and liquids, leading to the birth of ancestral spirits as children, as 'It is an *ale*, child *utokwobe*, coming out [birth] due to *tomya's totote-ale? maje*, winds returning back from far away place'. It is an *mejegeteebe* [distinctive from the term *enegetebe* which means embrace]. The *tomya* becoming *ale* is *akwan-geneegabe?* [reincarnation (in the case of human body)/recycle (in the case of animals)/regeneration (in the case of plants and trees).]

periodically over the bones so that the bone ornaments do not become too cold (due to *keye?*) and continue to emit smell. Once the bones are tied with *keye?* and covered with red clay paint, string made from plant fibres is tied to them along with shells and glass beads, so that the bones can be worn on the body.[16] The bone ornaments are kept within the family, packed in small baskets. At the time of moving from one place to another the baskets are also carried along. Some of the bone ornaments are also kept in the forest tied to a tree near the place of burial. Only on particular occasions like sickness are these bone ornaments taken out of the baskets and worn by individuals. They are also given occasionally to the *torale*, spirit communicator, for establishing contact with the spirit of a dead person. However, among all the recovered bones of a dead person, the lower jawbone is a special case. At the feast given after the burial, the chief mourner wears the ornament made from the lower jawbone. This ornament, known as *ibeedange*, is worn till the following full moon and then, like other bone ornaments, stored away by the family.

Ibeedange is the word for a lower jawbone that has the teeth intact. However, semantically *ibeedange* is composed of two terms: *ibee*, which means bad smell (stink, which brings danger), and *dange*, which means body (including the human body, and the bodies of plants, animals, and wood). Some other aspects of Ongee culture are also referred to as *ibeedange*, and these need to be considered here to understand the full meaning of *ibeedange* as the term for the lower jawbone ornament. The Ongees also refer to the part where the adze blade (the cutting edge, identical to the lower teeth) is tied to the handle of the adze as *ibeedange*. *Toneyage* (adze) is made of wood (from a plant, i.e. *dange*) and held together by tying and binding an edge which cuts. The two crucial aspects of binding and cutting thus come together in the *ibeedange* of the *toneyage*. In the same way, the canoe made out of wood by means of *toneyage* is also called a *dange*, since it is a prime object made from a tree. The front edge of the canoe is regarded as the cutting edge of the canoe as it moves through the water. This 'cutting' of the canoe,

---

[16] *See* Radcliffe-Brown, 1964: 126, 184; also Man, 1883: 182, plate IX, for examples of bone ornaments.

Toneyage
The adze used for
shaping wood. The part
where the splitting blade
is tied to the
handle is called
Ibeedange

Ibeedange
Ornament made out of an
ancestor's lower jaw-bone.

The front of canoe
(Dange) that cuts through
the water is referred to
as Ibeedange.

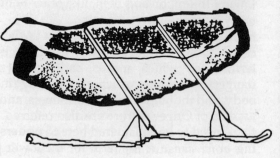

Pig hunting arrow.
The part where the arrow
point is connected to
the shaft is called
Ibeedange
(This is also true for the
turtle hunting harpoon.)

Fig. 9 Examples of Implements that Use the Term *Ibeedange* to Represent Identical Parts

and especially its front edge, is called *ibeedange* since in the canoe are held 'bound' Ongee, turtle, and dugong hunters. A good, hunter 'out at sea' is seen by the Ongees as well tied to the *ibeedange* of the canoe, and this quality makes the hunter capable of cutting the *dange* of the turtle and the dugong with a harpoon, and of then recovering the hunted animal by 'binding' it down with the cord attached to the detachable harpoon point. The point where the arrow for hunting pig is attached to the main shaft of the arrow is also called *ibeedange*. The same is true for the harpoon poles, which have a connecting point that is called *ibeedange*. It is at the *ibeedange* of the harpoon that the detachable point for killing the turtle and the dugong are tied with long cords (*see* Fig. 9).

It is within this framework that the lower jawbone of human beings and of animals gets the name *ibeedange*, since it holds food, and to it is tied the 'cutting edge', i.e. teeth. It is thus at the level of conceptual meaning that cutting becomes a release of *ibee*, a dangerous smell outcome of the process involving the hunting and killing of another *dange*, such as the body of pig, turtle, or dugong. It is in hunting animals that canoe, harpoons, and arrows are used. In the case of the lower jawbone of humans and animals, it is the teeth which cut the food (derived from *dange*) of an animal and release bad smell, and it is the lower jaw that the Ongees therefore cover with white clay paint after consuming food.

The Ongees explain the short periods of time during which the bone ornaments are 'openly' used, and the process of packing them 'enclosed and covered' in a basket at home, as a means of avoiding the process of *otenduabe* (attraction). Kanchu, who showed me the bone ornaments for the first time, said:

> It is good to wear the bones of the dead person for a few days after they are made into ornaments. The smell released reminds the *tomya* of his *ibeedange* with his or her relatives and he should not *inacekame* (forget). It also tells the spirit that we are not *inora* (not far away). But then we have to keep (*ibeedange*) *inena* (far away) so that the *tomya* continues to look for it again and again so that we have the *ale utokwobe* (childbirth) . . . of course, when someone is sick, we take the *ibeedange* out and wear it so that spirits come down and help us. It is always to be carried by the *melame* or those who are

moving alone so that one's own *tomya* comes down to protect the individual against the bad spirit.

The principles of *malabuka* and *talabuka* constitute the rationale for the Ongee preservation of dead relatives' bones. The *ibeedange* ornament made out of the lower jawbone emits the smell which is released and then absorbed by the ancestral spirit. The *ibeedange* is the source of attraction between the jawless body of the ancestor spirit and the jawbone ornament possessed by the spirit's living relatives. The attraction of the ancestral spirit implies that the living Ongees will receive help, protection, and more *dange* in the form of children. The smell of the bones attracts the jawless spirit to the places where the Ongees are, where the spirit enters various fruits. When the women eat these fruits, the spirits trapped inside enter the woman's body. The spirits who are thus trapped and eaten become the *geerange* (bones) and the *gebo* (flesh) of a child's body. This leads to the Ongee belief that the bones of the dead attract, by means of smell, the boneless bodies of the spirits, and that this attraction starts the transformational process by which the boneless body becomes a child's body. The bones of the dead are 'socially valuable', since they aid in transforming spirits into living beings within the community.

The bones of dead animals (those hunted and killed by the Ongees) operate in the same manner as human bones. Consequently, the skulls and lower jawbones of pigs, turtles, and dugongs are carefully preserved by the Ongees and kept in their homes. At the end of each hunt and after cooking, the skulls of the slaughtered animals are carefully preserved. The skull is dried over smoke and is covered with red clay paint, but never with white paint. The animal skulls are stored above the cooking areas of the camp, and are believed to release the smell of those animals. The smell of the dead animal's skull is dispersed and causes living animals in the forest to be

attracted to the place from where the smell is originating. The smell from the skulls of the hunted animals causes the animals to be hunted to remain in the place where the previous animals were hunted, a place where Ongee can go and hunt them and bring them back. If the skulls of the hunted animals are not kept then the living animals realize the death of animals and think of being hunted

. . . so the animal would leave the place where we Ongees found
them previously.

The dispersal of the smell of the dead animals is further aided
by painting the skulls red, red paint being conducive to the
release of smell, and by hanging the bones near the cooking fire
so that the smell is carried by the winds to the places where the
other animals are still alive.

Implications and Analysis of Gobolagnane: *Some General Considerations*

The *gobolagnane*, described as socially-valuable and magically-
potent things by Radcliffe-Brown, are related to the Ongees'
basic patterns of existence: a life that is much based, founded,
and dependent on moving around and hunting. For the Ongees, this life is also intricately woven and fabricated in relation
to the movements of animals and to those of the spirits around
them. The *gobolagnane* make it possible for the Ongees to move
around in relation to the animals they hunt and the products
they gather from the places around them. The *gobolagnane* also
make it possible for the Ongees to move around in relation to
the movement of the spirits, who share and coexist in space
with the Ongees. This sharing of space with the spirits is an
essential reality of which an Ongee must be aware and recon-
cile with. In certain ways the ideas about luck, success and fate,
life and death, forming as they do what we define as the
constituents of a world-view, are given concrete expression
and practice by the Ongees in their use of the *gobolagnane*.

Fire and smoke, along with clay paints and bones, are
gobolagnane, that make it possible for the Ongees to safely move
through various places. The *gobolagnane* make it possible for
the Ongees to 'go and bring', in other words, move about to
hunt and gather. The movement is made possible and, moreover, safe by the *gobolagnane* because they are instruments that
affect and control the movement of smell. In Ongee life, acts of
hunting form the basis for the relation between men, spirits,
and animals. The aspects related to hunting are explained and
embodied in the ideas of *malabuka* and *talabuka*, but throughout
their life, to hunt and not be hunted depends on the movement

of smell. In light of the act of hunting, that the spirits and the Ongees are related through the process of birth or death makes the relationship between them self-reciprocating. Every time human beings make way for the spirits within the shared space, childbirth takes place within the community.

When the Ongees refrain from being present in the place where spirits are hunting and gathering, and restrict the movement of smell by using *gobolagnane*, they ensure that there is no death. At the same time, by provisions within the shared space, the Ongees also make the birth of children possible. By letting the spirits hunt, the Ongees prevent themselves from being hunted, and this reciprocal arrangement enables the Ongee community to experience childbirth. The Ongee describe the situation of not being where the spirits are hunting, and preventing themselves and their smell from moving to where the spirits are, as 'not having a *malabuka!*' Once the spirits have left a place the Ongees hunt and gather there. This hunting and gathering, where the Ongees give food (a form of *talabuka*), is a means by which the spirit trapped in the food is transferred and transformed into a child in the womb. When a *malabuka* is avoided, what becomes possible is a *talabuka*. When the Ongees refrain from moving into a place occupied by spirits, and keep their smell restricted, childbirth becomes possible. Childbirth in the community is an impact experienced and created by the act of avoided movement and restricted smell. Consequently, the avoidance of a *malabuka* on the part of the Ongee actor not only has an effect on the man–spirit relationship but, more importantly, the Ongee experiences the impact of that act in the form of a *talabuka*. But failure in avoiding *malabuka* too has an impact, and one which is twofold. The Ongee community experiences the death of an individual, and there is an increase in the community of the spirits because the dead person is transformed into a bad spirit.

Consequently, the Ongee notions of *malabuka* and *talabuka*, both embedded in the act of hunting, involve and affect two distinct identities and actors, that is, Ongee humans and Ongee spirits. Every act, whether it is a *malabuka* or a *talabuka*, has two impacts, and they are reciprocal in nature. If an Ongee is not hunted by a spirit, then the spirit community is open to a potential loss, because a spirit may remain trapped in food and may get transformed into a child. The reciprocal situation occurs when the spirits succeed in hunting an Ongee, and a loss is experienced by the Ongee community, while the spirits experience an addition to their community. This reciprocity of act and the impact sets up the asymmetrical relationship between the Ongee and the spirits. It is in this relationship that every act of the Ongees has an impact not only on the spirits but also on the Ongee community.

The asymmetrical relation characterized by reciprocity is maintained between those who act by means of, and those who experience the impact of, the *gobolagnane*. It is the *gobolagnane* that make it possible to maintain asymmetricity in the reciprocal actions between the Ongee and the spirits. It is the *gobolagnane* that make it possible to avoid a *malabuka* and create a *talabuka*.

The *gobolagnane* help to indicate the absence and presence of the Ongees as well as of their smells, and this function is described by the word *gukwelonone*, game of hide and seek. When the spirits become aware of the presence of smell and of the Ongees they are able to hunt and kill human beings, whose death is a form of absence. The absence of the Ongee and his smell from where the spirits are creates the potential presence of human beings, since this is the situation in which spirits are transformed into children. Therefore, the preferred and guiding principle for Ongee actions and for their use of *gobolagnane* is that one's presence causes another's absence. This makes the hunting world of the Ongees alternativistic, whereby human beings and spirits share space, but the humans try to act in such a way that the reciprocal impact is of a positive kind and is an alternative to the negative kind. The Ongees therefore try to seek every opportunity to avoid a *malabuka* and to have a *talabuka*.

In this opportunistic alternativistic world of the Ongee, the *gobolagnane* are 'efficacious', since they index the self-reciprocating, asymmetrical relationship. The *gobolagnane* make men and spirits in the game of *gukwelonone* start as asymmetrical players, of giving and receiving smell, but provide men with all the opportunity to remain in an advantageous position.

The *gobolagnane* effect the movements of men, animals, and spirits by controlling the movement of smells. Radcliffe-Brown

was right in referring to the 'nature of . . . symbolic representation of the forces that affect the social life' (Radcliffe-Brown,
1922: 311) within the Andamanese as 'olfactory-sensations' and
'smells' (ibid.). On the basis of our consideration of the *kwayabe*,
smell, from the Ongee point of view, Radcliffe-Brown was not
far from the mark. However, 'smell' cannot be regarded as a
'symbolic representation of the forces' that affect 'social life'
when in fact it constitutes the social itself. In other words,
'smell' not only represents something but is itself represented
in terms of the use of the *gobolagnane*. It is smell that defines life
as well as death and makes real human beings the transmitters
of smell and the spirits the receivers of smell. Smell constitutes
the social, since it is the link between 'being acted upon'
(*malabuka*) and 'acting upon' (*talabuka*).

The Ongees describe *kwayabe* (smell) as something which
moves in the same manner as the tides of the sea. In the Ongee
language the tides are referred to as *kwayaye*, *kwaya* being the
common root verb in both the terms for the tides and for smell.
Kwaya as a verb also represents the presence of one of two
entities. Thus *kwaya* in the Ongee language is used for expressing an idea like 'either this or that'. The Ongees explain the
notion of 'either this or that' as *gwayamkaio?be*, where *gwayame*
means high tide and exhaling breath. The second part of the
word is *kwaio?be*, which means low tide and inhaling. The
Ongees explain that it is never possible for low tide and high
tide to happen simultaneously, just as it is never possible to
inhale and exhale at the same time. Only one of the two is
possible. However, one follows the other. Within this framework of explanation, the Ongees also put men and spirits. That
is, at a given moment in a given place, either men hunt and
gather or the spirits hunt and gather. What is posited, the
identity of breathing, tides, and presence of man or spirit, is
that the simultaneous presence of two, a situation contradicting
the notion of hide-and-seek, causes the breakdown of order and
the creation of chaos. For instance, the dugongs are seen to be
a creation of the movements of low tide and high tide fused
together (Myth no. 4). In the same way, the myth of *durru* (Myth
no. 5) presents the case of men and spirits being present in the
same place, and the chaotic, disorderly impact of the transgres-
sion of *gwayamkaio?be*. *Gwayamkaio?be* thus is the description

of the quality of relationship between man and spirit that generates the self-reciprocating asymmetric aspect. Asymmetric
relations are set up on the premise that spirits as bodies do not
have bones and therefore do not release smell. Spirits only
receive and absorb smell. Human beings, *dange*, living things
with bones, emit and release smell. Indeed, the ability to emit
smell distinguishes the living from the non-living. Humans can
emit as well as receive smell. Spirits can only receive smell but
not emit it. In view of the classification of living and non-living
bodies, men and spirits, and the principle of *gwayamkaio?be*,
gobolagnane are a kind of barrier between smell-releasing human bodies and smell-receiving spirit bodies. When the Ongees use *gobolagnane*, human beings acquire the spirit quality
of not releasing smell. Therefore, the use of the *gobolagnane* sets
up a symmetry within the existing assymmetical relationship
between man and spirit. It is through this capacity of the
Ongees of using the *gobolagnane*, that the humans and spirits
are in accord with the principle *gwayamkao?be*, that men remain
men and spirits remain spirits. Since *gobolagnane* also prevent
a *malabuka* and make possible a *talabuka*, the addition of new
human beings and spirits also continues. In the absence of
gobolagnane the asymmetry changes to symmetry between men
and spirits. This would imply that within a given place the
gwayamkao?be aspect will fall apart—creating disorder and
chaos, as in the mythological explanation of dugong (Myth
no. 4) and that of *durru* (Myth no. 5). If the symmetric aspect is
not set between men and spirits, the sequence of childbirths
would come to an end and one by one all the Ongees would be
hunted by the spirits.

   In the course of my field-work I enquired from Ongees if
they had any explanation for the significant decrease in the size
of the Ongee community on the island of Little Andaman.
Often the Ongees would remain quiet and become sad but on
being asked again and again they said, 'No body is gone, all the
spirits and Ongees are still here. Only after the spirits' anger
the Ongees become less and less. Then our smell was not good
enough—not affected enough by the *gobolagnane*—more and
more spirits hunted Ongees—the bad spirits have become
many!'

   Some of the older people feel that one day all Ongees will

become spirits and then once again the ancestral spirits will come down to the island and if no wars take place then things will be back to a position of sustained *gwayamkaio?be*.

In relation to the models of *malabuka* and *talabuka*, the *gobolagnane* and principle of *gwayamkaio?be* reveal aspects of olfactory-kinesics. This idea of smell and its movement is central to Ongee society.[17] The movement of smell is controlled and related to other possible forms of movement, going much beyond the purview of just moving from place to place. The control of the movement of smell is the conceptualization of the movement of life into the domain of death and, conversely, the movement of death into the domain of life. Thus smell is to be 'tied' and 'cut' to maintain the game of *gukwelonone*, hide and seek, where each move is intended to become absent for those seeking presence or those seeking the hidden present. That is, men hiding from the spirits and men themselves seeking the hiding animals are all embodied in the model of hunting.

Torale: *A Case of Exception to the Use of* Gobolagnane

Avoiding contact with the spirits is significant for all the Ongees and for this *gobolagnane* are used by all. All the Ongees are in the same position with respect to the spirits. The only exception to this are the individuals who are known as *torale*. Within the Ongee community the *torale* have a distinct position of influence.[18] They are capable of not using the *gobolagnane* and yet establishing contact with the spirits, and thus function as spirit communicators who create *malabuka* with the spirits without the help of *gobolagnane*. Of course, anyone could

[17] In the course of daily coversations Ongees ask after each other's life and well-being by inquiring: '*Konyune? onorange-tanka?*'
This means literally, 'when/why/where is the nose to be?', in other words, this is the Ongee equivalent to the English, 'How is it going?', 'How are you?' or 'What brings you here?' When an Ongee points to himself as 'me', he puts a finger at the tip of his nose. In everyday conversation, an Ongee first places his finger on his nose to index himself, and then goes on to state the purpose of his visit.
Spirits, in day-to-day conversations, are also referred to in terms of the nose. Spirits are called 'long-nosed' things, i.e things whose noses precede the sensation of or presence of their actual limbs and bodies.
[18] I use the term 'influence' as used by Radcliffe-Brown (1964: 47) to distinguish it from notions like authority.

abstain from using the *gobolagnane* and yet contact the spirits, but what makes the *torale* 'influential' is that their contact with a spirit does not end in death. The *torale* are capable of having an *enegeteebe* with the spirits and coming back to the community of the living Ongees. This makes the *torale* the only individuals among the Ongees who occasionally contact the spirits, move with the spirits, and return to human society. The two *torale*, Koyra and Muroi, on Little Andaman are believed by the rest of the Ongees to be individuals without whom it would be difficult to coexist on the island with the spirits. It is said,

> If the *torale* leaves the island and becomes the spirit forever then we all shall become spirits—no one would be alive and even those who stay here on the island will know nothing about the seasons and spirits' arrival and we will not know what to do when.

The *torale* as individuals in the Ongee community are identical to what Radcliffe-Brown called an '*Oku-Jumu*', a dreamer, a person who functions as the 'medicine man' within the Andaman Islanders' society (Radcliffe-Brown, 1964: 48, 51, 176, 186, 301). Muroi and Koyra are believed to have influence in terms of their unique power and position of being *torale*.

The *torale* are responsible for extracting *ebebarro*, knowledge, sight/vision, which is capable of inducing heat. This sight possessed by them is 'extracted' by them from the spirits. For the *torale* to gain knowledge from the spirits a special movement is undertaken known as *metakabe*, a flight along with the spirits up to the sky, away from land and sea. The knowledge gained by the spirit communicator is a vision from 'above the forest'. To understand this notion, we examine a typical sequence of interactions between the *torale* and their fellow tribesmen and between the *torale* and the spirits.

Stage I

As one season comes to an end the Ongee visitors at a *torale*'s home slowly empty their little cane baskets. They take out lumps of red clay, the neatly wrapped up bones of their ancestors, handfuls of betel-nuts, blunt arrowheads, and adze blades. As everybody empties out their baskets to leave the contents with the *torale* amidst the sound of betel-nuts being cracked and the occasional popping of the smouldering fire, the Ongees begin a chant-like request. It may be just an oral performance, but it is a speech which has come down over a long period of time, ever since the Ongees and their *tomya* (spirits) started sharing and moving around in the same space. The Ongees chant:

> We are not finding any food, we are getting loose from our tie within this place, and the place we have been moving in! We feel pain, even the betel-nuts cannot subdue the hunger. Nothing cuts! It's all becoming dry and falling apart . . . we are no more heavy . . . here are the bone ornaments from our homes. Go go go along with the winds . . . would you above the forest see our parents and children; perhaps the *tomya* can give something to us!

After such requests are made the *torale* generally gets scraps of iron or old arrowheads (never new ones), and people give him lumps of red clay, *chendange*, the seeds of a creeper belonging to the family of *Ipomoea presceprai*, along with the betel-nuts. Ongees regard it to be important to give iron pieces, betel-nuts, and *chendange* to the *torale*. Acceptance implies that he will undertake the trip, an expedition, a risk of establishing contact with the spirits. For the *torale*'s *malabuka* it is essential that he has the red lumps of clay so that he can heat up his body and lose all the smell from his body so that the spirits do not follow him back to the residential area. Iron scraps are given to the *torale* to enable him to offer it to the spirits as gifts 'from children'. Gifts of iron are the most preferred items for the spirits since the spirits can make them into *maonole*, tools, which are essential for them because of their inability to cut things and masticate and bite into them for lack of a lower jawbone. Betel-nuts are regarded to be items of *coge* (food, as a category), which on consumption does not satisfy hunger, since they are not like other 'soft' foods. However, the hardness of betel-nuts makes them much like *dange*, which contributes to the formation of a strong body and 'suppresses hunger' while kept in the mouth. *Chendange* is regarded to be a spirit repellent. In case all the spirits who come for the *enegeteebe* turn out to be 'bad-tempered' and uncooperative, the *torale* is supposed to throw the *chendange* into a nearby fire. When thrown the *chendange* produces a loud noise which also succeeds in pushing back the winds and this is believed to scare the spirits.

Stage II

A few days pass and the reque$ters have brought more red clay, ancestors' bones, betel-nuts, arrowheads, and cutting blades to the *torale*'s shelter. It is now certain that he will go on a journey, but no one knows when. The *torale* tells me:

> I have to go . . . I will go, I alone can go to the forest, I will go to my own *megeyabarrota*. I need to take some fire without any smoke. I need to collect all the *alame* (red clay paint) I have got. I should not forget the iron, betel-nuts, and fat of animal. Just where I have my *macekwe*'s (ancestor's) hut I will keep the fire. Then I will go to the *berale* and take all the bones including the *ibeedange* which others have given me. After spreading the bones on leaves I will sit with my head placed in-between the knees and close my eyes and go to sleep. As my body inside *eneteea* tries going out of the body *matee* (body outside that-contains the 'body inside'), I will start dreaming. My *eneteea* will then collect all the smell I have given away in various places. While the *eneteea* tries to come back the spirits come and put the *eneteea* into *matee* and tie me up like a pig after it has been hunted. I will feel very cold at this tying up of my body by the spirits. I have to then hold the iron I have in one hand, realize that *akwebwekete* (dream) is over. My *eneteea* has not come into *matee*, which generally happens to all of us when we get up after sleep. Then a lot of spirits collected due to the *ibeedange* I have with me. I met my own *macekwe*. I then feel very cold . . . I have *dobolobolobe*. While I am feeling *lololobe* all the *tomya* start feeling my body. They are blind so they put hands and feet all over my body . . . I shiver and feel more cold. With so many spirits touching me like a blind person I become very light. All my body liquid is sucked up; I feel very dry and cold.

Stage III

For three days Muroi, the *torale*, was not to be found. All of us knew that he had gone to communicate with the spirits, but no one would really talk about him. His absence, and the taboo against referring to him in his absence, made him like one who had died. However, the immediate relatives of Muroi had strips of pandanus leaves tied on their bodies, to mark an occasion when they were expecting the arrival of someone who was not just anybody but the master-craftsman, influential Muroi. Similar strips of leaf are worn by members of a family in which a woman is pregnant.

Stage IV

On the fourth day Muroi returned with a basketful of sedimentary stone chips and bone ornaments, and told me:

> All the spirits and my own ancestral spirits came to me. Just like the helpless pig after it has been cut with the arrows is carried on the back, the spirits put me on their back. All the spirits want to do it so I get tossed around, everybody is sucking my body liquids, I am becoming light—very light. Finally I start my flight upwards, going up and down like the snake on the ground, I am finally above the forest. I can see all the forest, the pigs in it, all the sea with the turtles in it, all the trees with the cicadas and honey in them, all the creeks with the crabs and fish in them, and all the ground under which is the tubers and 'sharpening stones'. It is all to be seen from above the forest. Above the forest I see it all—men below have to look for it because all is hidden from them . . . On reaching above the forest the spirits take me to where the sky is . . . then they all start talking among themselves and start untying me. The good spirits are looking upon me and the bad ones decide to throw me into hot stones and water. I start vomiting, so much I vomit that inside of me seems to be turning out. The good spirits embrace me very strongly. At this point I have to start telling the good spirits that I have brought iron for all of them. The iron bits are then distributed by the good spirits among the bad spirits. The good spirits who are related to me tell the bad ones that I should be allowed to go back because I am their own child and now that the iron is with the spirits my jawbones will be no great addition. After a long debate the spirits come to an agreement and the bad spirits leave me. Now the good spirits tell me that I have to make sure that the Ongee, all children should behave and leave some honey and cicada for the spirits to eat . . . I have to agree and remember all they ask for themselves, when they will come and from where and what all should be left on the platforms at the horizon for them . . . the good spirits then leave me and throw me down. They choose a place where I should fall down. When I fall down there is a loud noise, and I crash, my skin is all covered with bloody injuries. It hurts . . . this happens because when I fall, I crash into the ground at the place selected by the spirits. It is at this place that a lot of *tejage* are shattered and upturned. I collected the *tejage* (sharpening stones) so that they could be brought back to the community. This is essential for making new the iron *maonole* (tools) so that cutting, binding, and living is possible . . . I then warm up my body near the fire I had left, cover my body with clay paints and start my walk

back home. After reaching home I take many baths and sleep for a long time . . . Once women tell me that I do not smell, I break the *tejage* I had brought along with me and distribute it to all the people around me. I tell them all that I saw while I was above the forest. I tell them about the spirits I saw and also what the good spirits were planning to do. It is important that I remember all that I saw and heard from the spirits and keep remembering it till my next *enegeteebe* with the spirits!

Here the good and the bad spirits refer to the distinct classes of spirits, differing on the basis of the nature of death of an Ongee and the resulting dead body from it. Ongees regard 'malevolent' spirits in day-to-day conversation as bad spirits. Ongees who die in accidents and their bones are not retained by the living relatives within the community of the living Ongees become bad and malevolent spirits. In the description of spirits, given by the *torale*, the good spirits refer to those spirits whose *ibeedange* are used by the *torale* to establish contact.

The *torale*'s position of 'influence' in the Ongee community is the result of both his capacity as a knowledgeable person, *ebebarro*, and that he,and he alone, is capable of acquiring knowledge by not using the *gobolagnane* when he establishes contact with the *tomya* (spirits). The *torale* establishes contact with the spirits. When he undertakes this movements, he is not 'going and bringing back smell', an idea that underlies the use of clay paints, bones, and smoke. Instead, the *torale* submits to the spirits, by letting them be attracted to his smell, and instead of dying, returns with 'knowledge', *ebebarro*. When the *torale* makes a move without the *gobolagnane*, a conscious act on his part, he permits the spirits as actors to act upon him (ref. to Stage II).

The fact that the spirits are allowed to act upon the body of the *torale* is an act itself on the part of the *torale* himself (ref. to Stage IV). In other words, the *torale*, unlike others, does not avoid *malabuka*. He, indeed, fosters a situation in which he has a *malabuka* with the spirits. It could be said that the power of the *torale*, which makes him different from other individuals and enables him to gain a position of influence, lies in that he, and he alone, is in a situation, by effort and not by chance, in which he permits the spirits to 'hunt' him. The *torale*'s *malabuka* is also distinct, because, in getting himself hunted by the spirits, he does not get hunted completely. While others hunted by the spirits do not return, the *torale* comes back alive as the same individual.

The *torale*'s return from the 'place of the spirits' is a mark of distinction and makes him distinguished, since he comes back a more knowledgeable person. The going of the *torale* to the forest and his meeting with the spirits are of great importance to the Ongees. This is evident in the content of the request the Ongees make to the *torale* (ref. to Stage I). It is evident that the Ongees are referring to their inability to find what they are looking for; in other words, their game of 'hide and seek' is not yielding 'good luck' in the form of a *talabuka*. Thus, individuals report *ulate*, failure, in operations involving cutting, a characteristic of the hunting and gathering activities. The 'request' also is a form of inquiry about the change of season, indexed by the loss of humidity and the falling apart associated with death.

The *torale*'s *metakabe*, flight upward along with the spirits, is of even greater importance to the Ongee community. The Ongees believe that the *torale*'s *malabuka* and *enegeteebe* make it possible for them to live and continue life—expressed by the phrase *yeenehye?-bagabeh*, which literally means to cut and continue living. It may not be wrong to say that, for the Ongees, to live implies cutting successfully (ref. Stage I). The spirits, characterized by their lack of a lower jawbone, are believed to depend on the Ongee for gaining cutting power. Also, this makes them especially fond of such spirit food as cicada grubbies (*tombowage*) and honey (*tanja*), food which can satisfy hunger without mastication. Because of this the *torale* carries gifts of iron, which can be used by the spirits to make cutting implements and tools. Further, because the spirits lack cutting capacity, because they are always hungry, and because 'spirit-food' like honey and cicada are not available all the year round, the *torale* carries the gift of betel-nuts for the spirits. However, when the *torale* visits the spirits, the iron pieces he gives them become the prime objects from which cutting implements are made. In return, the *torale* comes back with *tejage*, bits of sedimentary rocks which the Ongees use to sharpen the cutting blades they have. The *torale* is regarded not only as making the further cutting operations successful for the Ongees, in terms of the *tejage* he distributes, but also as bringing back 'information' about the future movements of the spirits. He knows which spirits of which kind will come to the land of the Ongees and eat what kind of food, and is able to say how, when, and where the Ongees should move in relation to this movement of the spirits. Thus, the Ongees believe that the *torale* alone, after his *enegeteebe* with the spirits, can generate further potential of the cutting edges to cut, and can guide them to the places where cutting would be possible. All this has to happen within a space that is inhabited by the Ongees as well as the spirits; both have to continue living together, sharing the resources.

Both the *torale*, Muroi and Koyra, reported that it is important not to forget what they learned from the spirits and saw while moving upwards from the forest along with the spirits. According to the *torale* themselves, if they forget, they have to go again, and it is believed that it is not good to visit the spirits often, since this weakens the visitor's body. However, it is important that every now and again the *torale* must go and pay a visit to the community of spirits. The visit is also good for the spirits, since it is seen by the Ongees as a means of providing the former with the objects of *yeenehye?-bagabeh*, for living and cutting, i.e. food and pieces of iron.

It is evident that in his relationship with the spirits the *torale* puts his life in danger. He gives to the spirits, especially bad ones, what they need, i.e. means to live and cut, and also negotiates ways of providing food for the good spirits. When he enters into a giving relation with the spirits, he also brings to the community of human beings the perspective of 'above the forest', which is actually the perspective of the spirits, although the spirits cannot see everything but smell it all from their position above the forest. The *torale* also brings *tejage*, sharpening stones, of whose location only the spirits know. The *torale* thus becomes like a spirit, though he remains human. He breaks the asymmetrical relationship between men and spirits.

The *torale*'s distinct position of 'influence', created by his capacity to face danger and to be of special use to the community, is highlighted by *torale*'s three characteristics, which he does not share with any other individuals. The only exception is that, after his death, his apprentice may demonstrate the same power and skill in establishing contact, gaining knowledge, and transmitting it.[19]

The three characteristics are:

1. The *torale* does not use *gobolagnane* but uses the smell released from his body to attract the spirits and to have an *enegeteebe* with them. He does not avoid *malabuka*, but rather enters a situation whereby a *malabuka* becomes possible. In this event, the *torale* permits the spirits to act upon him and his body.

2. The *torale* uses his capacity to create coincidence to gain not only contact with the spirits, but also information for the whole community. He thus helps his fellow tribesmen in organizing their actions and movements in relation to the movements of the spirits.

3. All the individuals who have an *enegeteebe* with the spirits do not come back alive. A person will die after having an *enegeteebe* because he cannot signal by means of his smell for assistance from spirits. Even if an individual makes it to the 'sky' with the spirits, he can never see what the *torale* can. Never can a non-*torale* succeed in negotiating between the good and bad spirits. The *torale* alone is an individual who has the capacity to have a *malabuka* without dying, and to come back with information that helps the whole community to continue having *talabuka* (conjunctions) and successfully avoid *malabuka* (coincidence).

These three characteristics of the *torale*, differ from each other in terms of the strategies of smell brought into play, and the relations *vis-à-vis* the spirits, but they replicate the same basic principle. This principle is demonstrated in the way Ongee use *gobolagnane*. The *torale*, by not using the *gobolagnane*,

---

[19] The importance of the *torale* in the community's life is evident in the desire of every Ongee parent to have a *torale* as the *mutarandee* to his or her children, especially the male children. To become a *mutarandee* means that the individual becomes an initiator as well as the individual who adopts the children. Consequently, the majority of the male children are attached as the *mutarandees* of one or another *torale*. The *mutarandee* relation means that the individual so entered will be trained to be a hunter. However, the *torale* only teaches the art of doing *torale* to his sister's son. If the *torale* has no nephew, he will accept another male child of the community as his apprentice in learning the skills of doing *torale*.

purposefully creates a *malabuka*, and in the process gains, so to say, information which the other Ongees use to avoid a *malabuka* and to create a *talabuka*. In other words, the *torale* is responsible for telling the Ongees about the changing winds, different seasons, movements of the spirits, the different food resources, where they are to be found, and when they are to be consumed. It would be right to consider the *torale* as not only 'the dreamer', as Radcliffe-Brown (1964: 176) describes him, but also the Ongee individual who is knowledgeable, knows about meteorology, topography, anemology, different patterns of movements, and spirits. It is the *torale* who makes it possible for the Ongees to coexist with the spirits. It is the knowledge of gaining the knowledge of the *torale* that enables the Ongees to move along with the shifting winds and the spirits. Information given by the *torale* influences the actions of the Ongees, whereby a *malabuka* is avoided and a *talabuka*, in terms of finding different things at different times, becomes possible. When the *torale* allows himself to be partially hunted by the spirits he makes it possible for the other Ongees not to get hunted by spirits and to continue hunting.

In not using the *gobolagnane* and in creating a *malabuka*, the *torale* demonstrates the asymmetrical relationship between man and spirit. However, the outcome of the *torale*'s *malabuka* is the way to systematically and effectively use the *gobolagnane*, whereby further *malabuka* are possibly avoided not just by means of the *gobolagnane*, but also by means of the information given about where the spirits would be at which time of the year, thus making *talabuka* possible. The Ongees' use of the *gobolagnane* and the *torale*'s not using the *gobolagnane*, both have the same result in the following chronological events.

The way the *gobolagnane* are used by the Ongee at large and are not used by the *torale* produces the same desired effect in terms of first, avoided *malabuka* and second, created *talabuka*. This presents an identity with Radcliffe-Brown's description of substances like beeswax (Radcliffe-Brown, 1964: 152-3,156-7), which on being burned starts storms and can stop the storms. Radcliffe-Brown cites accounts from E.H.Man's work on the islands and regards the storms as a form of anger caused by spirits (Puluga and Biliku). References to a substance like burning beeswax that generates the storms and brings an end to the storms are identical to the Ongee *gobolagnane*. *Gobolagnane*, though remaining constant, produce different results. The *torale* in not using *gobolagnane* has different results from all other Ongees who use it.

Radcliffe-Brown's consideration and analysis of a substance like burning beeswax, give no weightage to who performs the act of burning and when it is done. He confuses the source of power with the agent of power. Radcliffe-Brown's (1964: 302) interpretation of Andamanese culture shows that power is possessed by the spirits and contact with spirits is dangerous. However, in certain circumstances it may benefit the society. For Radcliffe-Brown (1964: 306-7), Andaman Islanders' social life is a 'process of complex interaction of powers or forces present in the society itself'. Nowhere in Radcliffe-Brown's work do we find consideration and explanation of act, the actors, and substance and how the same act on the same substance produces different results. Ongees' use of *gobolagnane* and the *torale*'s use of it show that act and substance produce different results whereby the 'complex interaction of power' may be differentiated.

The question is, is it the use of the substance, is it the act, or is it the context of relations established by means of the substance and the act, that alters the relations? What is powerful and uncontrolled becomes controlled by the powerless. The usage, transaction, and exchange between the powerless and the powerful (an asymmetrical relation) creates power and makes it available to the powerless. Thus, in the accounts of Radcliffe-Brown, storms that are uncontrolled represent anger of the spirits and are controlled by the 'medicine-man' using the same substance and activity. How is it that he is capable of doing it? Does he enter into some power relation in the course of the activity of controlling storms (*see* Radcliffe-Brown, 1964: 152-3, 157-78)? This is reflected in the fact that the Ongee *torale* does not use the substance to create *malabuka*, whereby he gains some form of *enakyu?la*, power, from the spirits (exhibited in his giving of things to the spirits and bringing back some things to his fellow men), and makes it possible for others to avoid a *malabuka* and to create a *talabuka*.

Considering the use of the *gobolagnane* as the means of controlling the movement of smell, we have two cases. In the first, the substances are used by all to avoid an encounter with the spirits. In the second, the *gobolagnane* are used with the explicit desire to establish contact with the spirits. Here we see that the same *gobolagnane* are used to produce two different results—one of avoiding a *malabuka* with the spirits, and the other of having a *malabuka*. This is identical to the substances and acts that offend the spirits and lead to the formation of violent storms. On the other hand, the same substances and acts can control and stop the stormy wind conditions. This brings us to the problem of the power in these activities that have identical substances and acts producing different results. From the analysis of the *torale*'s usage of the *gobolagnane* and all the Ongees using the *gobolagnane*, it is clear that there is a self-reciprocating asymmetrical relationship between man and spirit. Both forms of the result are potentially within the substances, just as in the Ongee logic of *gwayamkaio?be*. The *torale*'s action (of not using the *gobolagnane*) and the day-to-day context (of using materials) suggests that the end result is an acknowledgement of the capacity to maintain the self-reciprocating asymmetrical relationship between men and spirits.

However, the position of the *torale* is such that he enters into a relation with the spirits by using the *gobolagnane*, which makes him akin to a dead person, enabling him to see what the spirits can perceive and to know about the spirits, and all this makes his character much like that of a *tomya*. If this is so, then in the *torale*'s performance he makes himself powerful, and the power is produced and externalized when he becomes like the *tomya*. Thus, the formulation could be reduced to the level of a simple proposition for an inquiry into the power of ritual as follows: the powerless person gains power by submitting to the powerful, and once he becomes powerful the power remains with him for further use. In this process there is a point where the asymmetry between the powerless and the powerful is first changed into asymmetry, and then the powerless gains power from the powerful by becoming identical with the powerful one himself.

In terms of the first question raised by the observations of Radcliffe-Brown, the 'medicine-man' alone can control the storms by doing the same activities which may start the storms because he is like or is one of the spirits, whose anger and pleasure are experienced by the islanders as the arrival of storms and the stopping of storms. It is valid to see the two questions in the light of Ongee culture, where it is important to have the *gikonetorroka*, ritual, for stopping the storms and winds by offending the spirits. This is essential because success in stopping the winds and in performing the ritual of *tanageru* gives the Ongees the power to start the next season, and the coming of the winds and spirits. Thus, the first offence is similar to the *torale*'s avoidance of the *gobolagnane* that stops the winds and helps the *torale* to establish contact with the spirits. This first 'event' leads to the other (reciprocating reflective event), where the relations are not asymmetric. In the case of the Ongee community at large, once the winds have been stopped, they can carry on with the initiation ceremony, and this is like the *torale* who, once in contact with the spirits, is in a position to gain 'perspective' and knowledge from the spirits. In both these cases, the powerless are made into the powerful. It is believed by the Ongees that both are a *gikonetorroka*, 'since it is power gained for the others', and because 'the power is brought back within the Ongees for cutting and living'. Just as with the return of the *torale*, the life of the Ongees, coexisting with the spirits, becomes possible and the end of the initiation ceremony makes it possible for the winds to start. In other words, the end of the *gikonetorroka* of the *tanageru* starts the winds again that were stopped before the initiation ceremony. Thus, both cases show the way power is generated and 'usurped' by the powerless, transforming him into a powerful person. The power, of course, is derived by affecting the asymmetrical relationship.

Granting Radcliffe-Brown's formulation that society has a 'sense of dependence' on 'power outside' (Radcliffe-Brown, 1965: 153-77), it is clear that the Ongees have to depend on their own movements, the movements of the spirits, the movements of the winds, and the movements of smell. But the 'power outside' is located in man, like the *torale* moving to the spirits— getting knowledge (power) from the spirits who are, so to say, 'outside' Ongee society. Since the spirits are 'outside' the asymmetrical relation of power is demonstrated in the Ongee concern for *malabuka*, *talabuka* and the *gobolagnane*.

Gobolagnane, as substance, effects the forms of events (*malabuka* and *talabuka*) in terms of the contact they establish or contact not established. The nature of the events is actually characterized by the relations which are established by and through the substances. As the contact established by the *torale* demonstrates, the effort to come into contact with powerful spirits in itself makes him a powerful, influential person, in terms of having knowledge. This could be extended to the Ongee context of rituals pertaining to stopping of winds and initiating young boys.[20] The power is created, extracted, and acquired by being in contact with the powerful. Powerful and powerless are both consistent, based on the principles of *gwayamkaio?be*, but in the presence of one the other is absent. In the process of gaining power the powerless become powerful through a shift of identity constituted of elements like being absent to being present, being hunted to becoming hunter, and smell-releasing body to smell-absorbing body. The shifts are made possible through *gobolagnane* between humans and spirits who are bound by the principle of *gwayamkaio?be*.

---

[20] I will be considering mainly the ritual of 'stopping winds' and initiation of the boys because unlike some other rituals within the Ongee society these, from the Ongee point of view, are regarded as particularly 'bad-offensive acts since they deal with the aspects of offending spirits' (Radcliffe-Brown, 1964). They are regarded as being distinct from other rituals in Ongee culture in that initiating and stopping winds deal with contact with the spirits who in day-to-day life, through *gobolagnane*, are demonstrated to have power over Ongees.

CHAPTER 5

Structuring the Structure of Ritual

Introduction

Ritual (*gikonetorroka*) activities between the months of early March and late July mark the duration that is anemologically significant since, in early March, through the ritual of *getankare*, Ongee space is emptied of winds and spirits. After the performance of the ritual, which is primarily *beti* (offensive in character), the Ongees undertake the collection of honey. In the months following May, the Ongees assemble and conduct the ritual of *tanageru* (initiation of boys). This initiation ritual is also characterized by the Ongees as offensive in nature and as an *akwabeybeti* (bad-work), since it causes the departure of the spirits and the absence 'of the winds'. It is during *tanageru* that angry Dare comes to the island for a short stay. The cessation of the anger of all the spirits is marked by the change of season from Torale to Kwalakangne? around late July.

Rituals performed in Torale end the spirit-given seasons and bring in the man-made seasons. Around late July, when the bad and offensive initiation ritual comes to a close, the man-made seasons also end and the spirit-given seasons are ushered in. From the Ongee point of view, the rituals that frame the duration of the man-made seasons (March–July) have a shared characteristic. Both rituals are 'bad-work' and are 'offensive', and both have an impact on winds and spirits associated with the seasons. In the beginning (early March), a ritual causes the spirits and winds to leave the island. In the end (late July), a ritual causes the return of the winds and spirits.

For the Ongees the rituals are a means of *tamale* (making). Analysis of the rituals that frame the duration between early March and late July show how the Ongees 'make' individual human bodies. It is this making of an individual human body, through a series of transformations and processes, which gives the relations between initiator and initiate and spirits and winds cultural importance to the Ongees. Rituals are of such significance in terms of 'making' that the rituals in the given duration are considered not only to affect the individual initiate, the collectivity of the initiators, and the spirits, but also to create a distinct phase of man-made seasons within the cycle of spirit-given seasons. It is because of the rituals and their effects, not only in terms of achieving initiation but also in expelling and ushering in the spirits and winds, that this ritual phase can be examined for aspects of the power of rituals. Within Ongee culture it is through rituals that power becomes an exchanged entity between spirits and human beings. This framework for analysis and consideration also posits that power effectively resides at two levels: first, with the spirits and, second, with the Ongee.

The effective distribution of power among the Ongees and between the Ongees and the spirits is exhibited in the use of *gobolagnane*, the *torale*'s journey, and the models of *malabuka* and *talabuka*. Since power is distributed between the Ongees and the spirits, the act of one affects the other, in terms of having an impact that alters various events of life through which man and spirit share space. The outcome of this power relation between man and spirit is the formation of the Ongee seasonal cycle. Within the seasonal cycle, spirit-given seasons are brought to an end through rituals, and this starts the duration of the man-made seasons. The man-made seasons also end because of a ritual that enables the spirit-given seasons to start again.

Winds, which move anywhere and everywhere, carry smell that affects the power relation between the Ongees and the spirits in terms of who hunts and who gets hunted. Winds and spirits moving together through places cause the Ongees to experience a change of season, which sets off yet another form of movement, since the food resources move and so do the Ongees to hunt and collect them. The movement of the spirits is indexed by the presence of storms and winds; indeed, the spirits and the winds are related to each other, and the presence of spirits is marked by the movement of the winds.

When the Ongees undertake rituals they expel the spirits, and along with this the flow of winds is restricted. When the Ongees perform another ritual to usher back the spirits, the flow of the winds also re-starts. This makes it possible for the Ongees to create and end a seasonal duration by means of rituals.

Within the seasonal duration created by the Ongees, their human society undertakes to endow certain rights and a certain power to the initiates. It is only after the Ongees have started the man-made season that the *naratakwange* (novice) becomes a fully initiated male member of the Ongee society. Thus the creation of the man-made season is simultaneous with society's act of changing a novice into an initiated person. The season and the life of an individual are therefore both outcomes of spirit movement into a place inhabited by the Ongees. Consequently, the seasons and an individual life (in terms of childbirth) are given to the Ongees by the spirits, but in the specific duration marked by ritual the seasons and an individual life (in terms of initiation) are made by the Ongees themselves.

The creation of a man-made seasons sets up the stage for the play of power between man and spirit, when men act and spirits suffer the impact. Once the stage is set, as I shall show in the following descriptions of the rituals, men become spirits in relation to the initiate. The man-made seasons in which the Ongees conduct the initiation ritual show that the power distributed between men and spirits oscillates, just as the *gobolag-nane* mediate between man and spirit and exemplify the effect of power between man and spirit in terms of *malabuka* and *talabuka*. Consequently, when the Ongees make a seasonal duration (like the spirits) and transform a novice into an initiate (just as the spirits transform themselves into children), it becomes clear that not only is the power of one over the other subject to variation, but is also a factor which makes men and spirits interdependent on each other (*galawelatetaye*), like the interdependent fingers of the hand.

The interdependence between man and spirit, possibly due to power shifts, connects the two types of seasons (man-made and spirit given) with the Ongee belief and value that men should not hunt and gather where the spirits are hunting and gathering. This characteristic interdependence, therefore, makes it possible for man and spirit to move together

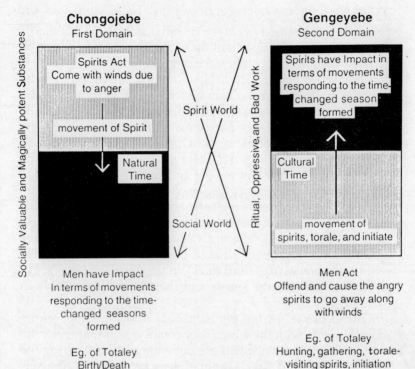

Fig. 10 Schematic Representation of the Two Domains

through the places in a shared space without clashing. This principle is expressed in Ongee culture through the notions of *malabuka* and *talabuka*, where there are two distinct forms of activity, men hunting and men being hunted.

The act–impact relationship, founded on aspects of the movement of smell and of wind, forms the diarchy within the Ongee world. Within the shared space, all *totaley* (activity) falls under the two domains *chongojebe* and *gengeyebe*. Here the notion of domain refers to the Ongee idea that when man shares space with the spirits, there are two levels of relations through which what Ongees get and what *tomya* (spirits) get differs by the two distinct *igagame* (time). For the Ongees the two domains are forms of *injube* (space) in which man and spirit interact. In the two types of space, *chongojebe* and *gengeyebe*, *igagame* (time) itself is different. *Chongojebe* is a 'domain' where man has no control over what he may obtain. Everything, i.e. *choge* and *maonole* (food and tools), in the domain of *chongojebe* is first available to the spirits and then becomes available to men.

For example, within the *chongojebe* domain, when spirits move in along with the winds they create seasons whereby the men have to eat what spirits are not eating, characterizing the Mayakangne? and Kwalakangne? seasons in terms of avail-ability of pigs and turtles. *Totaley* (activity) constitutes the basic act–impact relationship, and the two domains are distin-guished in terms of the different forms of interactions. It is *chongojebe* when the spirits are present (thereby causing the spirit-given season), and the acts of the spirits have a strong impact on men. When the spirits are forced to be absent from a place by the performance of an Ongee ritual, which is offen-sive in nature, men create a seasonal phase which is man-made. This man-made season exemplifies a duration when the do-main (the elements present in space for *totaley*) shows that men act and the spirits endure the impact. (*See* Fig. 10 for schematic representation of the domains in which different kinds of *totaley* take place.)

Different forms of *totaley* reflecting the characteristic re-lationship between man and spirit differ in relation to the domain. That is, in *chongojebe* the spirits act and men suffer the impact, which leads to *totaley*, in which the Ongees are controlled and affected by the spirits, and in response they restrict the movement of body and smell. This aspect of controlling, affecting, and restricting is referred to as *gekalakwebe*.[1] The domain of *gengeyebe* is an inversion of *chongojebe*, since men act and the spirits are controlled in terms of restriction of movement, an effective outcome of the Ongees offensive rituals. Thus, in the two domains, the *gekalakwebe* changes, just as in the two models of *talabuka* and *malabuka* the prey changes.

In each domain the one (man or spirit) who acts and causes the other to experience a reaction, is referred to as *gekalakwebe*. The one who suffers the impact, in terms of being controlled, affected and restricted, is referred to as *gugekwene*.[2]

If we extend the two types of domains, where different *totaley* take place due to change of *gekalakwebe* (the controller, the one who acts) and *gugekwene* (the controlled, the one who suffers the impact), the seasonal cycle and its constitutive seasonal durations also get divided up. Mayakangne? and Kwalakangne? fall under the description of the *chongojebe* domain, whereas Dare and Torale, which are characterized by men causing the spirits to exit and enter *injube* (space) come under the *gengeyebe* domain. Therefore, at the end of the month of March when Torale, the season of honey collection, sets in, the Ongees say, 'It is now *gengeyebe injube* we make the spirits do all the *gugekwene*!'

Both men and spirits are subjects of a shared space within which they move and are interdependent. In addition to being subjects of a shared space, men and spirits are also subject to a relationship of control which changes because of man's ritual activity. Once a ritual is performed, who the *gekalakwebe* is and who the *gugekwene* changes. This change of relationship and role between men and spirits is possible because 'time' itself is different in the two domains. A time when spirits cause all the men to respond and move accordingly forms, so to say, the 'natural season' characterizing the *chongojebe* domain. Within

[1] *Gekalakwebe* is literally the causative form of 'making something to happen'. Thus, the actor in the domain is *gekalakwebe*.

[2] Since spirits can control the availability of the food and tools available to man, spirits have an impact on man which is referred to as *gayewekebe* (tying), *comowekebe* (cutting), or *gakwekene* (pressing). These three activities are collectively referred to as *gugekwene* (meaning being controlled in relation to [or by] the spirits).

the *gengeyebe* domain the men move in such a way that it affects the spirits' movement out of the places constituting space. This causes the spirits to look for and do what men did in the 'natural time'. Thus, the domain of *gengeyebe* is characterized by the movement that makes an altogether different *igagame*—a 'cultural season'. A natural season is given to men by the spirits, but the cultural season is created by men. If we regard the seasons as temporal durations, then it becomes clear that the Ongee world-view has two kinds of time—one that is given and one that is made. However, both kinds of time, natural and cultural seasons, are characterized and formed by movement within the same space. It is the coming in of the winds and the spirits that forms the domain of *chongojebe*, and the spirit-given (natural) seasons. It is the departure of the spirits, caused by man, which forms the *gengeyebe* domain and man-made (cultural) season.

Bringing about a change of *gugekwene* and *gekalakwebe* makes the seasonal durations different. The rituals that frame the man-made seasons within the spirit-given seasons thus bring about change in the domain and in the relations of *totaley* within the domain. The rituals of *getankare* (around February/March) affect the spirits along with the winds, causing them to leave the island. This effect of the ritual starts the man-made season that forms the *gengeyebe* domain. Around July/August, after the Ongees have performed the ritual of *tanageru*, the spirits along with the winds are ushered back into the island, and this starts the duration of the spirit-given seasons and the domain becomes *chongojebe*.

The domain in terms of space (i.e. the island) remains constant in both the man-made and spirit-given seasons. What is different is the experienced *igagame*, temporal duration within the *injube* (space). It is the performance of *gikonetorroka* (ritual) that makes it possible to experience different *igagame* in the same *injube*, consequently forming two distinct domains (*chongejebe* and *gengeyebe*), and two different relations between spirits and men (*gugekwene* and *gekalakwebe*) that exist along with two different forms of seasons (man-made and spirit-given) to constitute one whole seasonal cycle (*monatan-dunamey*). This makes *gikonetorroka* (ritual) a means not only of creating the shift of power between the spirits and the Ongees but also sustaining diarchy at various levels (such as the relations in relation to seasons, relations of work, and relations of place in space). Ritual, therefore, is the process of sustaining and generating diarchy, and of affecting the relationship between men and spirits.

Within this framework of analysis the Ongee *gikonetorroka* (rituals) are much like the *gobolagnane*. The *gobolagnane* can determine the outcome of an interaction between man and spirit in terms of *malabuka* and *talabuka*, which are different temporal experiences for an individual, in terms of being a hunter and being hunted. The *gobolagnane* create a different temporal experience for an Ongee individual by affecting the movement of smell and altering the relations between men and spirits within a given place. *Gikonetorroka*, rituals, can make it possible to experience two different temporal durations for the whole Ongee community within space by affecting the power distributed between men and spirits and by affecting the movement of the spirits.

Meaning of Gikonetorroka *and* Detababe

Gikonetorroka, the Ongee word for referring to ritual, is formed from three words combined together to form one verb. The word is constituted by three units, where *koney* means light, the prefix *gi* refers to discarding, and the suffix *torroka* means away from. Thus, the literal meaning, which also forms the Ongee idea of what ritual is: 'to discard away lightness'. The performance of ritual, in the Ongee language, is glossed by the verb *tamale*, which means to bring together, to build, to fabricate. *Totaley* is the Ongee word for doing work, but the actual phase, a temporal duration when work is going on involving operations like tying, cutting, and pressing, is called *tamale*.

When the Ongees have to say that they are involved in conducting a ritual, the idea is expressed as '*Le tamale gikonetorroka!*' meaning, 'We [are] constructing [a phase, a body, a process] to discard lightness away'. The Ongees are very specific and clear about the idea that a ritual, in terms of doing away with lightness, is a phase, a process focusing upon one or more certain *dange* (human body). In the *tamale* of *gikonetorroka*, a specific human body or several specific human bodies are made light by a specific body or specific bodies. The body or bodies that make other bodies light, themselves become *babeye,* heavy. The idea implicit here is that in the process of ritual the weight is transferred from or transferred by the ritual participants. The one who becomes the focus of the ritual suffers a loss of weight, and those who perform the ritual become heavy. Becoming heavy by being involved in making another body light is referred to as *detababe.* Therefore, since ritual focuses upon shifting weight (making light and becoming heavy), it is much like work in a domain of *chongojebe* or *gengeyebe,* in which one acts and someone else suffers the impact of the act. Furthermore, ritual and work have a structural similarity with the *gobolagnane* which, in its effect on smell, determines who hunts and who gets hunted.

For the Ongees, performing a ritual always implies a *detababe,* and having a *detababe* always implies a *gikonetorroka* (ritual), just as doing work implies that one person acts and another suffers the impact, or that one controls and the other is controlled. Moreover, it is like movement, where one hunts and the other gets hunted.

The body which is light because of excessive loss of smell and liquids attracts the spirits, and it is easy for the spirits to take such bodies away. It is this aspect that makes the light-weight body particularly dangerous.[3] The process of making a body light in the course of ritual, therefore, characterizes the danger faced by the person whose body is ritualized as well as those who conduct the ritual. This aspect of lightness and being in danger is applicable to the body of the initiate in the course of *tanageru.* The state of danger involved in *tanageru* is further intensified by the consideration that the spirits are angered by offensive ritual acts.

The Ongees believe that once the initiate enters into relations with others in the community for his initiation ceremony, the novice is not just a human body. The novice is a *naratak-wange,* a chambered nautilus shell. The novice is in an abnormal state since the upper half of his body (waist upwards) is heavy and the lower half is light. This state has to be altered within the novice's body. The process of redistributing weight within

[3] Thus, polite greetings on meeting an Ongee are communications and inquiries about the state of heaviness and lightness.

the initiate's body is called *tamale,* and is a transformation whereby the top half becomes *olo* (light), and the lower half becomes *geduba* (heavy). This transformation process requires other bodies, that is, initiators. The transformative act of the initiators is referred to as *goke?* and implies that in the process of making the initiate's lower half *geduba* (heavy), they themselves become heavy. This makes the ritual of *tanageru* a process of not only transforming the initiate but also of transforming the initiators. The initiators, in making the initiate light (*goke?*), themselves incorporate the weight of and from the initiate. Consequently, as an individual goes through *gikonetorroka,* all the other individuals involved in that ritual go through *detababe.* This places *gikonetorroka* (discarding weight/making light) and *detababe* (acquiring weight/ making heavy) in a self-reciprocating relation of act and impact.

The Ongees attach great importance to performing *gekonetorroka* as well as being involved in *detababe.*[4] The reason for this belief is that in the process of performing *gekonetorroka,* the initiate experiences loss of weight but also brings an end to the man-made season. When the man-made season ends, the spirit-given season starts, and this makes it possible for the initiators (who are subject to *detababe*) to continue hunting and gathering along with the spirits. Unless men and spirits coexist, the resources of the Ongee world would end forever, as is evident in Ongee mythology (Myth no.1, units A-E; Myth no.2, unit T).

Tambolae, the father of one of the boys to be initiated while I was on the island described what *tanageru, gikonetorroka,* and *detababe?* through the focus on the body, do to the relation between the Ongees and the spirits. He said:

To do *tanageru* is essential—we must try to do at least one boy a

---

[4] The Ongee distinction of *gikonetorroka* and *detababe* is important since they do not confuse the two impacts. All people involved in bringing about a change (*gikonetorroka*) form the *detababe* which is a larger collectivity forming the idea of ceremony. In this ceremony (*detababe*) is the placement of the *gikonetorroka* which is ritual. It is because of this Ongee distinction of *detababe* (ceremony) and *gikonetorroka* (ritual) that makes it important to consider the body and the bodies in terms of actors and acted upon throughout the phase/season after the spirits are offended and then are offended again to end the man-made season and usher in the spirit-given season.

tanageru every seasonal cycle. However, to do *tanageru* we all have to come together. For doing *tanageru* it is to be the Dare. For getting the Dare in our *injube* we all have first to make Torale. Unless and until we do and make it all dry and get all the honey for ourselves and cause the spirits to be angry the season of Dare will not be there. We all are responsible for the happening of *tanageru*—In doing all this together not only do we make the initiate go through what he is supposed to go through *gemey?he* (manipulation and transformation of body weight) but we who are involved in the *detababe*, make ourselves heavy and safe and succeed in ending the danger and series of activities that have made the spirits angry.

Tambolae made this statement before the honey-gathering season of Torale started (middle of March). Tambolae went on to say:

Through the *chongojebe*, it is the *tomya* who control all of us in terms of the tools (*maonole*)and food (*choge*) we can get and lose. Once we do *torale*, *detababe* and then the *tanageru gikonetorroka* we are in *gengeyebe* where we Ongee can control what *maonole* and *choge tomya* can get or lose. In Ongee controlling what the spirits get we make *gengeyebe*. In *chongojebe*, we Ongee do all the *ebe-bebare* (smell-kept good [good work]) so that we have a defensive against *tomya*. In doing the *detababe* of Torale and *gikonetorroka* of *tanageru* we Ongee are doing *akwabeybeti* (bad work), causing the *gengeyebe*; the bad work of ours angers the *tomya* and we see that spirits move and get as we want them to. This is important because the child who comes to us because and through the spirits and can be taken away by the spirits is all part of *ebe-bebare* and *chongojebe*. In *akwabeybeti* and *gengeyebe* we Ongee send the Ongee to the spirits and then bring him down to our own *injube*.

Tambolae's statement points to the belief that when the Ongees do *gikonetorroka* and *detababe*, these transformations have an effect not only on the human bodies involved, but also create a distinct temporal experience within space. In Tambolae's reference to 'Ongee send Ongee to the spirits and they bring him down', the distinct quality of the *chongojebe* and *gengeyebe* domains is further elaborated. What the spirits do to the Ongees in the *chongojebe* domain, in terms of 'sending Ongee children' and taking away the Ongees is shifted to the *gengeyebe* domain. Through the manipulation of the body in the course of *gikonetorroka* and *detababe*, the Ongees do exactly what the spirits do. In the course of the initiation ritual (*tanageru*) the Ongees send the initiate to the spirits (identical to the spirits taking away the Ongees) and then bring him back (identical to the spirits sending down children). Consequently, through *gikonetorroka* and *detababe*, the Ongees cause the death and the birth of the initiate, which the spirits cause in the *chongojebe* domain.

From the foregoing statement and the ideas outlined above, a ritual, for the Ongees, is an activity which has four characteristic traits:

1. A ritual is a special time in which the relationship between men and spirits changes.
2. Doing a ritual implies deriving time or 'extracting' it from space by means of offensive and defensive 'constructs' which affect the movements of men and spirits.
3. By means of ritual, what the spirits do for and to men can be reversed; that is, men do to the spirits what the spirits have been doing to men.
4. A ritual endows men with the capacity to do what the spirits have done in the domain of *chongojebe*.

Seasonal Cycle and the Ritual Points Within the Cycle

Winds from the north-east start blowing into Ongee 'space' from the month of November, and mark the arrival and presence of the spirit Mayakangne?. Mayakangne? does not leave the Ongees until the beginning of March.

At the end of February, through the beginning of March, strong cyclonic conditions develop. This marks the end of Mayakangne? and the start of the ritual known as *getankare*. As the Ongee, living in the coastal area, experience the cyclones, they plan their forthcoming trip to the deeper forests. Since men do not go out hunting turtle, the women have to gather the fish to feed the camp. Women also have to gather the *keye?* (*Dendrobium*), and the stems of hibiscus for making thread. The two materials are then made into ornaments, which are essential parts of the *tanageru*'s ritual ornaments. Women also start making new grass aprons. At this point of the year all the women complain about how obstinate the men are, because the men love to spend all their time carving the *dange-ketulabe* (wooden phallus).[5] This carving of wood by men is regarded to be 'bad work', especially by the women who have to work hard gathering food. Women, consequently, are very busy during the first week of March, and constantly complain about the men being lazy. The women say that the men have started remembering honey and, therefore, they (the women) now have to spend all their time gathering the fish which have been washed up on the shoreline (*see* L.Cipriani, 1966: 116-18), and looking for the yellow skin of the orchid plant (*Dendrobium*) known as *keye?*.

The men along with the male children, on the other hand, are busy whittling *ketulabe* from wooden pieces that range from three inches to three feet in size. The men regard this whittling as an enjoyable activity. Though the women know that the men are all carving *dange-ketulabe*, a distance is maintained between the workplace of the men and the women. The men never visit the place where the women are involved in the weaving and tying operations involved in making the *tebogeru* (string and bark ornament) that adorns the initiate's body during the *tanageru* ritual. Just as the men do not visit the women while they weave and tie the *tebogeru*, the women do not visit the men while they cut and carve the *dange-ketulabe*.

Division of work along gender lines[6] also leads to the separation of workplaces within the camp-site in all the preparations for the forthcoming 'offences', i.e. the consumption of honey and the performance of the initiation ritual. As the people at the camp-sites continue to work, they plan the process of dismantling the shelters (*korale*). There is one common topic of conversation among all the adults. This revolves around joking and speculation about how many fruits and how much honey will be available in the forthcoming season of Torale, and how many children will 'come down' in the process

[5] Men along with the young boys are all involved in not only carving the wooden phalluses but also all the old *ukku* (wooden carved containers for honey) are scraped and cleaned. Some new ones are also carved at this time.

[6] The men being involved with carving wooden phallus and women working on the tying and binding at the start of the *gengeyebe* domain marks the inversion of standard divisions of work type associated with men and women. That is to say, in the *chongojebe* domain men do all the binding and tying and women do all the cutting.

of childbirth and how many boys will 'go up' in the process of *tanageru*. This talk causes the younger and unmarried boys and girls to giggle. The boys play pranks (*ocholo*), such as running away with the unfinished *ketulabe*. The boys with the *ketulabe* run towards the place where the mothers, aunts, and sisters are working. This causes quite a stir among the women. The display of the semi-carved wooden *ketulabe* held outstretched or on their shoulders by the dashing and bashful young boys raises shouts of jovial questions by the women. The women shout, '*Beti-ye Beti-ye?* (offensive,offensive)—who gave you this—don't you know it is a bad thing!' All the women join in this chorus and try to catch the boy so as to snatch away the wooden *ketulabe*. Usually the boys outrun the women and successfully return to the place where the men continue to work.

In this process of chasing and fleeing, questioning and teasing, the boys try to touch the women with the wooden *ketulabe*, and sometimes the women catch one of the them. If this happens, then the boys shout out '*Mutarandee Mutarandee!*' *Mutarandee* is a classificatory term for mother's brother. Sometimes *mutarandee* is an appointed term of reference. It refers to an individual man who agrees to adopt the child even before the child is born. This makes *mutarandee* a term of reference for one who has adopted you as a child. As the boys shout for *mutarandee*, all the men leave their work and run to get the boy from the women who may be holding him. At the sight of the men the women let go of the young boy who has been held hostage. As the boy is released, the men and boys again separate from the women and girls, and the work of carving and cutting continues. The game comes to an end as the *ketulabe* are all carved and polished, covered with dabs of red clay paint and stored at the *obonaley* (residence of the bachelors).

As one by one every adult male finishes making the wooden phalluses the game of the boys and the women comes to an end. All the *ketulabe* are stocked up at the *obonaley* and the process of making the *ketulabe* and storing them comes to a formal end with *gobomamey* (full moon, literally, bleeding danger). *Gobomamey* is the night of high tide. Turbulent sea water rushes into the creeks and coastline. The madly racing and rising sea water rips apart all that grows on the coastal area. The coastal lines are subject to reshaping by the tides on every full moon, but the first full moon night after Mayakangne? is particularly significant. This is the night of the full moon when the coastline will be reshaped, but before this starts the shaping of all the wooden *ketulabe* by the men must be completed.

As the water level starts rising, the women begin dismantling the shelters. All the men and boys gather at the *obonaley* and take out the finished *ketulabe*. Everyone holds the carved objects in their hands and sit around a fire. Then the men holding the carved *ketulabe* start *gigabawe* (singing, causing attraction). As an individual recites a line, the chorus repeats it. The singing continues until the early hours of the morning. From the night of the full moon to dusk one hears a constant repetition of the following:

Yehye?—ethee-gobomamey.	Listen!—it is full moon,
gogogye? cheleme.	bleeding is the moon
Mey?na mey?na-ne kuttu kuaye	Full full in a big way are the creeks
Ethee garitabe nanchuge	now the place is wet
enekutata tamale tamalegee	together construct the construction
Ikatalabe	dryness
Beti beetinenegee	Offensive! Offensive
cherabe gayabattetenya?bogalange	coming soon is the Tenya-bogalange (form of cyclone)
Koyrabe naratakwange	floating the nautilus shell
koyrabe naratakwange	floating the nautilus shell
Ethee beteeikatalabe	now is the offending dryness
Enekutata gobabannonegi gejalange	together we bring the secreting
gejalange garitabe gukkiewe	secreting are the wet honeycombs.
guke?ye korale	bees home
Tamalenenegee engabe	constructing are the women
akwangeneegabe enekutata ott	recycling all the
wa tanja-eye	fruits and honey
Beti Beti gogogye?	offensive offensive is the bleeding
Bejaie ma ma bejaie!	rain no no rain
Ethee-ikatalabe	now is the dryness
ethee gobomame	now is the full moon
getankare wa-gejalange	ushering the ritual of *getankare* [associated with wooden phallus] secretions.

Just before daybreak, all the married women go down to the point where the waters of the creek and the sea meet. The water is no longer turbulent because the tides are low tides. All the women walk into the water. When the water reaches their waists they remove their *nakwenage* (grass aprons, also known as *eneyagegi*). Then the young unmarried girls are called to bring fresh, newly made *nakwenage*. The young girls walk up to a few steps behind the older women standing in the water and the new *nakwenage* are passed on.

After putting on the new grass aprons, the married women let the old ones float out on the sea. By this time the sun is above the horizon, and all the married women and unmarried girls walk out of the water. By then the men and boys have finished singing at the *obonaley*. This is the last phase of the ritual of *getankare*, which had started with the men carving and the women weaving and tying. As the men get up to leave the *obonaley* they stuff their mouths with areca nuts, betel-nut leaves, and powdered clam shells. Meanwhile, the women who have returned after the change of grass aprons, start dismantling the shelters by removing the thatch, the posts, and the sleeping platforms.

The men walk to where all the freshly bathed women are congregated and are dismantling the shelters. The women have assembled at what was the central ground (*wabe*) of the camp before the shelters were dismantled. The married men walk in front of the women in single-file formation, and the young boys follow behind. All men and boys have their bodies smeared with ash from the fire around which the singing was done. Everybody holds a *ketulabe* and one by one they deposit it in front of the women. This is done three times. In the first sequence of placing the *ketulabe*, the men who are married walk up to their respective spouses and leave the *ketulabe*. In the second round the men place the *ketulabe* in front of their classificatory sister-in-law. In the third and last sequence, young boys place the *ketulabe* in front of their fathers' sisters. Once all the men and boys have placed the *ketulabe* in front of the women, they start circling the *wabe* in single file. Though the women are all very quiet and serious, the men by contrast are all very jovial and playful, trying to outdo each other in making loud sighing sounds. They start the procession round the *wabe* by mimicking the gait of various animals, such as crabs, frogs, turtles, and pigs. Every individual exhibits a different choreographed movement, replicating the movement of some creature. Some of the married men who have yet to become fathers put on grass aprons and hold a turtle shell in front of their stomach and imitate the gait of pregnant women. Until the men start imitating pregnant women, every choreographed movement seems to be comical and farcical, but after the selected men, acting out the walk of 'pregnant women' with a turtle shell, complete the round of the *wabe*, every man changes his movement. The last few moments of going round the *wabe* depicts the highly stylized, slow, and deliberate movement pattern of the monitor lizard and the civet cat. Those who imitate the monitor lizard bend over completely and walk on all fours. As the 'lizards' move about, the 'cats' behind them and ahead of them stop to create more space between themselves and the 'lizards'. In contrast to the earlier movements, and to the movement of the monitor lizards, the cats move with greater agility but alternate between standing still and walking behind the lizards. At this moment, on asking the Ongees what this was, I was told that, 'We are telling all the Ongee and *tomya* that we should not forget about *ayuge* (monitor lizard) and *kekele* (civet cat)!' On further questioning, I got only one simple answer from all the camp-mates, 'Now the search of smell—the hide and seek between us and spirits —has to change!'

It was after a lapse of time that the women told me about how the lizard and the cat looked for the turtle eggs. This was the segment extracted by the Ongee women from the myth of Ongee origin in relation to Eneyagegi and Eneyabegi (Myth no. 2, units F, G, and H). According to all the Ongees, what the men had done in depicting various creatures' gait, pregnant women, and especially the walk of the monitor lizard and civet cat constituted the *onolabe*[7] dance (Myth no. 2, unit I).

When the men circle round the *wabe* as monitor lizards and civet cats, they bring the *onolabe* to an end. Now nobody is allowed to joke or to be jovial, and seriousness and an air of

---

[7] *Onolabe*, the Ongee term for dance, is constituted of two terms, *onola* and *wabe*. *Onola* means 'others' or 'beyond', and is used to convey the idea of 'above all'. *Wabe* is the root word for crying out a feeling. Thus, *onolabe*, consisting of *onola* and *wabe*, literally means 'above all, beyond others a cry'.

solemnity prevail in the total ethos of the camp-site. One by one, all the men who participated in the *onolabe* and had brought wooden *ketulabe*, pick up their carved wooden products. The men exit from the *wabe* with the *ketulabe*, and go to the place where the women had bathed and cast away their grass aprons. On reaching the water-line, all the *ketulabe* are thrown into the sea. As these float away from the creek towards the sea, the men spit out the red residue of the chewed betel-nut and leaves. As this is done, the young boys enter the water and dip the newly made *ukku* and cleaned up old *ukku* (honey containers) into the sea water and bring them back to the camp. This marks the end of the *getankare gikonetorroka*.

After the conclusion of the *getankare* ritual, all the men sigh, and remark on how tired they are and what hard work the ritual has been. Again the men gather at the *wabe* where all the women are waiting. As the men meet the women, the women say to them in a very sarcastic and taunting manner, 'You should not be lazy! Why did you all not get tired when you were making *beetikatalabe totale wa tamale*?' In the women's usage of the term *'beetikatalabe totale wa tamale'*, meaning 'offensive dryness causing work', what is being denoted and referred to is the carving of wooden phalluses and the whole ritual of *getankare*, which ends with carved phalluses being thrown into the sea. When the Ongee women avoid the term *ketulabe* (carved wooden phallus), what becomes evident is the intended purpose of the *getankare* ritual, that is, work which causes offensive dryness.

The men finish packing up all the household goods, such as tools, containers, ropes, poles, mats, and baskets, from the shelters that were dismantled by the women. There is not much to be packed in the household of the Ongees who translocate in relation to seasonal changes, but when the men do this they exhibit a reluctance to come to terms with the situation that they all have to move, one by one, to the forests in the interior of the island. They have to move away from the place of coastal residence where they had been camping. They will now set up residence in the forest (using mainly what has been dismantled and packed), where the trees are now loaded with honeycombs. Once the shelters are set up in the forests, the men will start cutting down trees to make canoes, and will start the everyday search for honeycombs.

The Ongees believe that to go and live in the forest and gather all the honey they can find is made possible by performing the ritual of *getankare*. The ritual, which involves carving the wooden *ketulabe* and casting them away after the night of the full moon in the middle of March, marks the end of the Mayakangne? season. All this is bad work, which stops the violent winds from the north-east and, above all, ritually offends the spirits, sends the spirits away from the Ongee place, thereby causing dryness. This dryness results because, when the offended spirits go away, the winds and the rain leave the island, 'just as children leave and move along with parents!' The Ongee action, an offensive ritual, thus causes dryness (*beeti katalabe totale wa tamale*), which characterizes the season made by the Ongees.

Why is the Getankare Ritual Offensive ?

The season of Torale, associated with the collection of honey, is also associated with the performance of the *getankare* ritual within the space shared by the Ongees and the spirits. Torale is the season in which honey, a prime food of the spirits, becomes available to the Ongees. The availability of honey for the Ongee ritual of *getankare* has a twofold outcome. First, the *getankare* starts the man-made season and, second, the spirits along with the winds and rain leave the island, since they are offended. But why is this ritual—with its wooden phalluses, its singing of songs, its particular choice of the imagery of dryness, secretion, and fruits, its imitations of the talk and choreographed movement of pregnancy, its playful and jovial interaction with women—considered an integral part of the seasonal cycle, and why does it precede the season in which the boys are initiated?

When I asked the Ongees why the *getankare* is of significance to them, and what characterizes it as offensive, they gave me various responses. Some younger boys and girls within the camps did not have any explanations. For them the ritual was just a 'play' (*ocholo*) that began after the cyclonic conditions of Mayakangne? had developed and which ended on the full moon. It was a play that preceded packing up and following their seniors to the camp in the forest.

The most common response from the older women was:

> When Mayakangne? is about to end, the spirits are made to go away for sure. All the turtles have gone to the north-east direction. Cyclones are about to come. Rains could continue but all the men do the bad work of throwing carved wood in the sea and telling the spirits about this which makes the space without rain. Since *injube* (space) is no more *gejalange* (secreting) we get all the honey in *injube* which is *ikatallabe* (dry). This *ikatallabe* makes *injube* full of wet honeycombs. If men do not do the bad work then rains and spirits continue to be in our *injube* and honeycombs will be dry. We will have no honey for us!

The older men, on the other hand, told me:

> If the space was not made dry and spirits were allowed to eat all the honey then we will not be visited by the angry *onkoboye?kwe* [term referring collectively to the spirits who enter the fruits and consequently cause the pregnancy of women]. So by eating all the honey in the forest after we have made the rains and wind along with the spirits go away the spirits are angry, offended (*beti*) and we can expect them to come down again full of anger and hunger, filling up our space with wetness and winds. This also makes it possible for the women to eat the fruits which still have the *onkoboye?kwe* left in it. Thus when we do the *getankare gikonetorroka* some of the spirits remain here, trapped in the fruits and then we collect all the honeycombs and the fruits. First we eat up all the honey so that the spirits remain in fruits and those who do escape go up and tell Tenyabogalange and Dare about our bad work.

From the Ongee point of view, doing the *getankare gikone-torroka* is a way to ensure the pregnancy of women. Teelai, the oldest man among the Ongee, reported:

> All things, including the child, woman and us grow because rains and winds enter and cause things, all forms of *dange*, to become either hard or soft. Honey is a different matter. It is not available to us if rains and winds continue. With the rains and the winds go the spirits come; they enter inside the honeycombs. Once spirits are inside the honeycombs they are no more hard outside and wet and soft inside. Thus we need honeycombs to have children but to eat the honeycombs we have to stop the winds and rains which come along. Honeycombs and cicada grubbies are the only things which are firm outside but are drippy and secretionary inside. And they

both get destroyed by the rains. Honeycombs and *ketulabe* (not the wooden ones) are identical. They both are firm on the outside but are capable of secreting. When we give the wooden phalluses away in the water we tell the spirits about what we are going to the forest for—to eat the honeycombs so that we can have more children. We need to eat the honeycombs. It is all this which we remind women and spirits and to do it we have to do the *getankare gikonetorroka*, where *dange-ketulabe* (wooden phalluses) are cast away.

The Ongees do not believe that the male sexual organ has any role to play *vis-à-vis* the conception of a child. The basic function of the husband is to copulate so that the winds that get blocked in the womb, because a spirit is changing into a child, remain comfortable. If a husband does not copulate with the wife, then the exhalations of the spirit inside the woman's belly would accumulate and make her very uncomfortable. Given this concept of conception, man's *ketulabe* is seen just as a means or an instrument of altering the wind condition within the pregnant woman. It is this instrumentality which makes the honeycomb identical with the male organ in terms of the prospects of childbirth. The association of the honeycomb with the male sexual organ at one level, and the association of honey with spirit food as well as with child-causing food at the second level (since the absence of honey keeps the *onkoboy?kwe* [collective name for spirits] confined to the fruits), is further clarified in Ongee mythology. In the myth of the women and the spirits what the offence is, is identified. The myth also relates how the *getankare gikonetorroka* involving the *dange-ketulabe* associated with the ritual of *tanageru*. It is the narrative of the first *tanageru* which forms the discourse providing the explanatory background for the Ongees performing an offensive ritual. It is in this myth that one can locate various relations between the Ongees and the spirits which are subject to change. There one finds a description of the ritual act for stopping winds and affecting the spirits. In the myth Ongees tell of the first *tanageru* as well as *getankare* ritual and explain the aspect of offence in it.

Myth No. 6

First Rituals and Offence of Tanageru

First the women came from the sea. Then came the men from the forest.

[See myth about origin of life on the island of the Ongee, Myth no.2.] Since women came to the land of Ongee before men they were closer and more friendly with the spirits, especially Tenyabogalange. It was the teaching and friendship with the spirits of women which enabled them to collect and gather. The spirits also gave women the ability to bind and tie things very well. Women on reaching a certain age would release blood at each full moon. This capacity to release blood at each full moon was also the arrangement and ability given through the spirits. Spirits and women were really close to each other. They understood each other's presence and needs very well. [Unit A.]

The capacity of women to bleed was a means for the women to maintain their body always in a state of appropriate distribution of heaviness and lightness within the body. This was important. So much so, that in women's bleeding the spirit and women could both coexist. So the Tenyabogalange decided, on knowing how women were doing, to send *dugueyge* (snake) and *morukwe* (dove) in a *naratakwange* (nautilus shell). The snake and the bird would come in the shell just like people go in a canoe. Bird and snake were to come to Ongee and see what they were all up to. Bird would lead the snake to all those who had started bleeding. Under the guidance of the bird, the snake would go along and consume all the blood released from the body of the women.

Women became more and more like the spirits. Bird and snake would go back to Tenyabogalange and tell about those who had stopped bleeding. It was all right because Tenyabogalange knew all and that the women were not bleeding since they had spirits living inside their bodies. The bleeding was not possible since the blood was being consumed by the spirit inside. When spirit was not inside the women, the women would pass the blood over through the snake, bird, and shell to Tenyabogalange. Since women were so close to the spirits they did not do any cutting where blood would flow or cause secretion leading to the loss of liquids. This would all mean loss of weight. All this is *akwabeybeti* (bad and offensive work). Because women did not do all this, they never went to hunt animals. They would only gather and would do only cutting of the human body causing the release of smell and blood. [Unit B.]

When women started living with men they told the men to cut things away so that the blood from the women's bodies would not stay back and cause the discomfort. Men's activity caused the release of smell and departing winds. The women lost blood and became dry. Snake, bird, and the shell came down but they had nothing to report to Tenyabogalange. It was all dry: women and dry place. No winds, no smell. They told Tenyabogalange about the men doing all the bad and offensive work. Tenyabogalange was very disgusted and so he decided that he was going to bring all the women, especially young ones who had yet to start bleeding and all those who were bleeding. Men were left with only women for whom a lot of bad work had been done by men. The women who were left behind had been married for a long time. No children were coming to the Ongee community. [Unit C.]

So the men decided to do something about it. They had done so much bad work that winds had stopped and no spirits were coming down with whom they could have a talk. So the only possibility was to send some Ongee to the spirits. Since winds were not around Ongee men could not use the smell to get the spirits to come and have a *malabuka* and *enegeteebe*. Finally, it was decided that when the bird and snake came they would kill them and put young and lightweight Ongee boys in the nautilus shell and let them float away to the spirit's residence. [Unit D.]

Bird and snake came and they were killed by starvation since men had killed all the pigs in the forest and the young boys were fed with the blood of the female pigs and female turtles. Other men made sure that none of the bones and meat of the pigs and turtles killed were visible to the bird, snake, or any spirits. The chambered nautilus shell was occupied by the young Ongee boys who were light like the bird and snake. Also, the blood on which the boys had been feeding on excessively made the boys smell too only of blood. The profusion of the smell of blood led the shell to mistake the boys for snake. Ongee boys sat inside the nautilus shell and other Ongees pushed them into the sea.[8] [Unit E.]

Young Ongee boys reached the residence of Dare and told her that without women on the island Ongee men were very sad. Men had nothing to hunt and they did not know how to gather. The Ongee men's complaint was conveyed to Tenyabogalange. Men promised the spirits that they would make sure that they would never do any form of 'bad work'. At the instructions of the spirits it was agreed upon by the men that they would make sure that women bled and never would the men try to comfort the women. It was made clear to the men that if men continue to do *ketukabe* (comfort the women)[9] then no children will be born. Men promised to the spirits that they would not do any bad work and that they would look after women in accordance with the spirits' instructions. [Unit F.]

[8] Ongees regard the *naratakwange* (chambered nautilus shell) to be endowed with a unique capacity. So much so, that it always accompanies those travelling by canoe. It is not only a drinking container but is the only shell, from the Ongee point of view, which can float in the sea. No other shell has that capacity.

[9] *Ketula* meaning phallus also refers to the function of the digging stick for yams. From this root term originates the Ongee term for copulation, *ketukabe*. I gloss *ketukabe* as 'comforting', as the act of man in relation to pregnant women is seen as comforting the spouse who is already pregnant.

All the women and men got together. They decided that at a certain time of the year the men would hunt and women would gather. This division of tasks was to be followed throughout the year. However, it was not possible to do this all the year round, especially it was not possible for the women to go gathering after they had eaten too many fruits and become pregnant. Also, when the *bulledange* (jackfruit) was around the only thing available for the Ongees to consume was honey. The sea had no turtles worth eating and also the pigs were too young. So men agreed to collect honey. [Unit G.]

However, men were not supposed to collect anything, they were only supposed to hunt. Women alone could not feed the Ongees once Maya-kangne? was about to leave. It was essential for men to go out when women went to collect *bulledange*. Men, therefore, found it was inevitable for them to go and collect honey from the honeycombs. For men to get all the honey and eat it was deceiving the spirits since honey was food for spirits only. So men decided that they would throw away the *ketulabe* and the spirits would think that men are out in the sea.

When men did cast away the *ketulabe* they thought they would convey to the spirits the idea that they [men] were out hunting in the sea and they were not 'comforting' women. This made it possible for the men to go and collect honeycombs. What was not realized by the Ongees was that honeycombs and *ketulabe* both were comforting to women in a way. Since honey made it possible for the women to conceive the child inside, and that meant bleeding stopped and men could then do all the comforting women needed since they had let the *dange-ketulabe* float in the sea water. [Unit H.]

After a few seasons the spirits realized that the honey was not available to them and that the Ongees were consuming it all and, on top of it all, children were being born to the Ongees. Men continued to 'comfort' the women and what they [spirits] thought as *ketulabe* was actually only *dange-ketulabe*, it only smelled like the Ongees because the *dange-ketulabe* was all covered in betel-nut spit so the wood's smell was under the spirit's smell. Spirits got very angry and they came down all at once and took away all the grown-up men. [Unit I.]

Now women were left all alone on the island. Women had the company of only very young boys. Women could not get any honey, any comfort, or any children. So the women decided that they would feed the young boys with just blood and send them along with the nautilus shell to establish contact with the spirits; just like the men had done previously. This was the only alternative. Since angry spirits had gone and there was no wind and spirits left—it was the only way women could have got in touch with the spirits. Since men had created a dry space where no spirits and winds were present, all because the consumption of honey and casting away of the *dange- ketulabe* had offended the spirits. Women had also stopped bleeding since they had made the spirits very disturbed and they had too much honey to eat along with the fruits. Since men were not around the women had no alternative but to eat what they could gather. [Unit J.]

Women then went to the sea and changed their grass aprons, came out, and then they all got together and started singing songs praising the pigs. The pigs were attracted by this singing. As the pigs came close, the women succeeded in killing them day after day and started feeding the blood to the young male children. This was the first *tanageru* done by the women. The young boys' bodies changed from young children. They started regressing into younger and younger forms. As the bodies of the young ones got stuffed with the blood, women continued to massage them and heat them so that all the body liquids would go below the waistline but not below that. The bodies became very light. Women then killed the snake and the bird and installed the young boys along with pigs in the nautilus shell and set them afloat. To the bodies of the young male children the women attached various strings so that they could pull back the children if they got scared while going to the spirits' residence in the nautilus shell. [Unit K.]

On reaching the home of the spirits, the boys explained how they had come up. How their body was changed and transformed by the women doing *tanageru*. All the boys were questioned by the spirits. Spirits wanted to know what activities were being done by men and women. The boys were honest and sincere in telling the spirits about what the women had done, about how the spirits were tricked by the Ongees. The boys promised that they would never again do such 'bad work'. However, neither did the boys know nor did the spirits inquire about why honey was not available. [Unit L.]

The spirits took pity on the Ongees. They decided to enter into a relation of 'give and take' with the Ongees. Tenyabogalange brought down all the men and boys and left instructions that the Ongees were expected to eat *teralu* (needlefish) and not fat pigs when they realized that Mayakangne? and turtles had all gone. Tenyabogalange wanted the Ongees to eat also some mangrove fruits (*daboja*) [fruit of *Bruguiera gymnorhiza*] but not pigs, because pigs were for the spirits to eat. If the men started killing and eating fat pigs then once again the men would be all taken away. Each seasonal cycle the Ongees were to give and take according to the instructions of the spirits. Nowhere in this negotiation for establishing the 'give and take' relation did the question of honey come up. [Unit M.]

So the Ongees continued to get all the honey and also get back the boys as well as the young children. This made the Ongees have *gengeyebe*, where men took everything from the spirits according to their own actions. Ongees made the spirits angry by taking all the honey and stopping winds. Then they could also start the winds by doing *tanageru* (and in this way also get young boys [trapped spirits] in the wombs of women). Since spirits did not negotiate about the honey, every seasonal cycle the angry spirits would take away the boys of *tanageru* for their offence of hunting the fat pigs and then after questioning the hunters of the fat pigs, they would be returned back. Thus, the Ongees kept on stopping the winds and then doing again the same kind of bad work started the spirits to visit island and have the winds as well as Maya-kangne? and Kwalakangne? follow after the *gengeyebe* of Torale and Dare. [Unit N.]

It is in relation to this account of events and the relationship into which the Ongees and the spirits had entered that, to this day, the Ongees continue to conduct the *gikonetorroka* rituals and to 'construct' *gengeyebe*. In each seasonal cycle the Ongee boys are initiated and they are supposed to fearlessly go to the spirits and return. When the initiate goes and comes back, the whole Ongee society derives benefit from the ritual of *tanageru*. This benefit is well articulated in the words of Muroie at Dugong Creek:

> Every seasonal cycle we must do the *tanageru*. However, to do the *tanageru* we must have a good Torale and then only is it possible to have *tanageru*. If we get no honey then it is a loss of doing the *tanageru* because if the boys do not return back then we have nothing. No boys coming back no children to be born, no honey eaten. It is all bad and a loss. Mayakangne? and Kwalakange? just follow each other—much rain and water, no dryness. Spirits take everything and we get nothing from them. Therefore each season we have good Torale, we know that one who is married could get to the spirits, if he hasn't gone through *tanageru*. It is good to be married and then have *tanageru*—since eating honey with wife is truly eating the fruits. All this is important because this also tells the younger Ongee how they have to live along with the spirits. In every *tanageru* there is a *benchapuka* (death for a short while)—a danger but in each *benchapuka* there is a lot of—many many *utok-wobe* (birth).

From mid-March till the month of May (the duration of the period of honey collection) all the Ongee families separate and scatter. Each family sets up a shelter (*korale*) in the section of the forest which is traditionally associated with it. In each *korale* married people reside with their young children (including adopted children). The residents of the *obonaley*, who include unmarried men and boys, also shift from the coast to the forest and set up separate individual shelters in the section of the forest associated within their father's band. All the young and old unmarried girls go to reside in the *korale* that is constructed in the section of the forest associated with their mother's band.

Once the shelters have been set up, the men undertake activities associated with the season of Torale. The day starts with the men gathering at the work-site for canoe carving. Making canoes takes much labour and time. First, huge trees at least four feet in radius are felled. Then the bark is stripped off and the inside is carved out. This work continues with groups of men taking turns at working on different stages of the canoe till noon. Every day, in the early hours of the evening, all the men disperse from the site of canoe-making and go out looking for honeycombs. Meanwhile, throughout the day women gather jack-fruits and crabs. At nightfall men and women return to their respective family shelters. A meal of jack-fruit and crabs along with honey is consumed by everyone in the different shelters. At night the Ongees work by the light of a resin torch for the forthcoming *tanageru* ritual. The men start shaping new bows and arrows, while the women try to finish the string ornaments. Sometimes, in the evening, people from neighbouring sections of the forest visit each other. Visitors generally discuss the availability of honey. Usually on such an occasion invitations are extended to visit each other and share portions of the collected honey. If the visitors decide to stay over for one or two days, depending on the supply of honey in the host's forest, a new *korale* for the visitors is set up.[10]

Around the end of May and the beginning of June, the season of Torale comes to an end. The Ongees associate the end of Torale with three distinct signs. The first sign is the disappearance of birds like *bele-bele* and *kentamale* (drongos, strikes, swallows, and martins) which guide the Ongees to the honey-

---

[10] Only in the duration of Torale are new shelters with green leaf thatch set up for the visitors. In all other seasons the visitor stays (if a man alone) in the *obonaley*, or, if a family visits, then the thatch of dried leaves is extended from the existing *korale* and the sleeping platform (*kame*) is extended for the visiting family to sleep on.

combs. The birds are regarded as friends of the Ongees since they feed on the bees which fall dead from the honeycombs after the Ongees have sprayed the combs with the chewed up leaves of the *tonjoghe* plant. The birds generally migrate from the Andaman Islands at this time of the year, and the Ongees see this as indexical of the honeycombs being exhausted. The second sign is the passing and formation of clouds at this time of the year as an end of the Torale, since they believe that the rainwater will enter the honeycombs (of *Apis dorosata*) left empty after honey collection. The third sign is the most important, and it is this sign that indicates the complete end of Torale. Around June, in the forest, the ground where the *gegi* tubers grow becomes all furrowed and upturned. This is the 'pigs digging work'. According to the Ongees, 'now pigs are frantically looking for *gegi*, since all the ripe jackfruit which fall from the tree have been collected by the women! Pigs need to feed their young ones so the pigs start digging up for *gegi* —and now the pigs are becoming fat and heavy'. Once these three signs are observed by the Ongees, Torale is said to be changing into *nakwarabe*. 'Forest is about to lose the dryness and become all wet', is the Ongee explanation for the end of Torale and the onset of *nakwarabe*.

It is now the time for the Ongees to end the act of *uye?uye* (stealing honey from the spirits), as soon the spirits start the process of *gengegetebey* (stealing from men). It is at this point within the seasonal cycle that the Ongees seem to be most aware of the relationship between themselves (non-spirits) and the spirits. It is referred to as *etetu-chegalemam-lebe*. Within the Ongee language the term *etetu-chegalemam-lebe* is the verb for picking up something with tongs without touching the heat or the food with the hands. The term is also used by the Ongees to express the relationship between themselves and the spirits, implying that, like tongs which make it possible for food to be picked up without really touching the fire or the food, men and spirit both 'steal' from each other, while moving from place to place without letting the other know that they have stolen all the honey and moved out. The movement of the Ongees and the spirits from place to place within the shared space is homologous to the folded strip of cane which functions as a pair of tongs (*etetu-chegalemam-lebe*). There are two distinct kinds of stealing in this manner, which characterize two distinct domains called *gengeyebe* (men act and spirits suffer the impact) and *chongojebe* (spirits act and men suffer the impact), respectively. In other words, within the domain of *chongojebe*, the spirits steal from the men (*gengegetebey*) and in the domain of the *gengeyebe*, men steal from the spirits (*uye?uye*).[11]

As the season of honey ends, when the forest becomes *nakwarabe*, the Ongees start leaving their forest residence. Elder couples are the first to leave the forest which has become steamy and hot because of the mild winds and clouds coming to Little Andaman from the south-east. One by one all the families set up new shelters at the place where the *getankare gikonetorroka* was held. The area where the water from the creeks merges with the sea, the location where the women discarded their grass aprons and the men threw away the carved wooden phalluses, is the site of the circular arrangement of the new *korale*. In the following days the men organize the activity of *ekyetakorakabe* (bringing out the unfinished canoes [canoes without outriggers] from the forest to the coast). At this camp-site, the families give each other the honey they have collected. This is an important *girorobuke* (gift), since it is believed that the consumption of honey indicates respect for each family's traditional section of the forest, from which the honey comes. Consequently, when the *girorobuke* honey is consumed, the Ongees say, 'Ah! the ancestors in your forest still have maintained good honeycombs!' As the gifts of honey are exchanged, the Ongees discuss the season of Torale in terms of all the work that was done. The discussion revolves around the praise and criticism of activities like *aaregurya* (making of new bows and arrows), *gotatebogeru* (making of string ornaments for the initiation ceremony), and *ganandare* (finding out if the pigs in various parts of the forest are fat or not). All the three activities, *aaregurya*, *gotatebogeru*, and *ganandare*, are important work and their outcomes make it possible for the Ongees to

---

[11] The word *uye?uye* is often used in the Ongee language as the adjective describing a person who looks around well when he is going somewhere, i.e. one who has good eyesight and a keen sense of vision. The word *gengegetebey* is also an adjective describing the quality of one who can see well without eyes, by smelling. In other words, one who has such a developed sense of smell that it even substitutes the faculty of vision (just like the spirits' keen sense of smell).

undertake the *tanageru* ritual. Even if all the conditions are right and everything is ready, holding the *tanageru* depends upon the availability of the uninitiated, young married man.[12]

Generally the *tanageru* should be held after every duration of Torale, but given the demographic situation of the Ongees attempts are now made to space out the frequency of the *tanageru* rituals and the number of individuals initiated at each *tanageru*. In the course of my stay among the Ongees of Little Andaman the *tanageru* was held after three complete seasonal cycles. Around the third week of May 1984, the Ongees at Dugong Creek decided to move from the creek to the coastal area. New shelters were set up. Each family's sleeping platform was set up under one common thatch and, instead of the circular arrangement of the *korale*, there were only two rows of sleeping platforms under two long thatch roofs.[13] The decision

[12] Ongees on Little Andaman reported that in the past it was possible to have *tanageru* every year, and that too with many young men at the same time—but now given the fact that very few of them are alive, marriage of the young ones does not take place for lack of women; everyone tries to select and plan in such a way that at least at the interval of two or three seasonal cycles they have someone to be initiated. In recent times there has been not only increases in the rate of decline of the total population but also the ratio of men and women has changed considerably. Many of the married couples are childless and there are more unmarried men in relation to unmarried women. Apart from historical reasons, such as contact with the outside world, the strong cultural preference, in fact a rule, is that one whose spouse is dead has the first preference for remarriage. This means Ongees are concerned that those who were once married should get married again as soon as possible, and they have an edge over the ones who are ready to marry but do not have a spouse. Thus, a lot of couples reflect a wide age difference and often the children within the family are from a previous marriage.

[13] In the central ground of the new camp a few *kame* are set up near the camp's fire. These are used on occasion of collective *gigabawe*, singing.

At the *korale* of the father whose son is to be initiated, a special *kame* higher than the one for sleeping is made. It is on this relatively higher *kame* that all the new bows and arrows, string ornaments, baskets, cutting blades which were made for *tanageru* in the course of Torale are stacked.

Generally, in an Ongee camp-site weapons and equipment such as bows and arrows along with new big cutting blades are never to be displayed. These things are generally tucked away in the thatch of the *korale* or in the forest so that one doesn't knows where the other's equipment is. On the way to hunting each individual goes on his own to bring the equipment, and then the group departs for hunting. This custom of hiding equipment is essential for all Ongees to follow for two reasons. First, if the *maonole* (collective term for all the bows and arrows and cutting blades) are near the *korale* then the *tomya* (spirits) may get it, and,

to move to the coastal area camp-site is the first community decision about holding the *tanageru*. The process of deciding to have the *tanageru* is started by the *manyube* (father-in-law) and *umoree*, father of the prospective novice. It is the *manyube* and *umoree* who first agree on having the *tanageru* and then present the proposition to the whole camp. If the majority of the camp-mates (including elder men and women) agree, then and only then is the *tanageru* held. After the decision to have Tambolai's and Totanage's sons initiated was made, the residents at the new camp-site made *eneyachuge* (swings). It is important for the Ongees to set up swings at this time of the year. All the children and women must swing so that the sensation of the winds touching the body, which may be forgotten by the end of the preceding season, may be remembered. Swings thus create a place within the place. The body sitting on the swing is in a place where 'wind' is a bodily experience, whereas the place around is devoid of any wind movement. Therefore, Ongees say, 'to be on swing is like wind. The body leaves smell from here and there just like wind takes the smell from here and there!'

Only children and women sit on the swings, and the experience is proscribed for men. The reason is that men during this time of year refrain from using clay paints, an important aspect of *gengeyebe*. Ottalatey said, while fixing up a swing for his children on the *wabe*, 'Now men go to the forest without any *gobolagnane*, in search of pigs—we are leaving body smell here and there in between forest and home! We do not sit on swings to leave smell, we move to leave smell!'

The men who start going to the forest to hunt pigs, without using clay paints, fire, and smoke, engage in purposeful, conscious action that attracts the spirits and also makes the hunting of pigs more difficult. The Ongees believe that, since they do not apply any clay paints, or carry any fire and smoke on this occasion, they make the pigs more protective about themselves. In other words, a *talabuka* becomes difficult. Not taking any measure to restrict the smell being dispersed from the body is also perceived by the Ongees as attracting the *tomya* to the place

---

if the *maonole* are easily visible around the *korale* then, in moments of anger, camp-mates may be tempted to use them or destroy them as an expression of anger

where the search for pigs is on. This is important, as the Ongees explain:

> All the *tomya* have gone—no wind but we want the *tomya* to come back and see the bad work of *tanageru*. We want them to know that we are going to hunt pigs and eat nothing else—so we have to *nateelabe* (invite spirits to see all this)—so no *gobolagnane* are used when going to forest. In this hunting *tomya* comes but pigs do not. Also the pigs are very brave in this season since they have young ones to protect and feed. We Ongees at the same place are also brave so that we can hunt brave pigs and also bravely bring back the winds and let the spirits see what we are doing—making them angry!

In the course of this pig hunting, the boys who are to be initiated are never taken along. The two important people who have to participate day after day are the father and the father-in-law of the boy to be initiated. Each day the team returns with a pig, if possible. The people in the camp congregate and examine it. One-by-one all the senior men and women come to the spot where the pig is placed (in the central ground of the camp). The pig is singed and then cut up into two standard parts, *gebo* (the meat) is piled up on one side and *gettange* (fatty skin-layer) on the other. All the senior men and women at this point make judgments about whether the pig has enough fat and is heavy enough or not. After two or three days of hunting by the father and the father-in-law, the people at the camp finally declare that 'The pigs are now fat enough and our hunger is increased—let us send the *naratakwange* (novice) to the forest—let the *tomya* see and hear us well'. This marks the point where the whole camp agrees upon the start of the *tanageru* ritual. When the pigs are no longer *dolakambey* (thin and lightweight) and are *idankuttu* (heavy and fat), the time comes for the appropriate pigs to be hunted by the novice, which forms the important central activity around which the Ongees organize themselves in 'constructing' the ritual of *tanageru*.

When the camp agrees that the pigs are *idankuttu,*the novice's father and father-in-law, along with the novice, visit each *korale* at the camp-site the following morning. The visitors place part of the *gebo* and the *getange* before each *korale*'s residents. This is a *nateelabe*, a form of extending an invitation to par-

ticipate in the ritual. If the residents of the *korale* take some of
the fatty skin of the pig, then they are regarded to be a family
who will come to the series of feasts held in the course of the
tanageru. Families who take only the *gebo*, non-fatty meat, by
virtue of refusing pig fat, are regarded to be the people who
will be involved in all stages of the *tanageru* as workers. Those
who take only lean meat are the 'feeders' of the *tanageru* ritual,
and accepters of pig fat the 'eaters'. On the following day
distant camp-sites are also visited by the father and father-in-
law. Each *korale* responds to the visiting father and father-in-
law by promising the things they will bring for the ritual,
especially what they have made in terms of tools, weapons, and
ornaments for the ritual. Often on this occasion, each *korale*
contributes at least a dog to participate in pig hunting and some
arrows. The party that had gone on the round of invitations
returns with the dogs and arrows. Those who accepted the
invitation in the form of only lean pig meat move close to the
residence of the person who is to be initiated. All those who
took the pig meat are now separated from those who accepted
pig fat.[14] The novice and his wife, the novice's father, and
father-in-law move to an adjacent area to set up new *korale*. The
new *korale* are close to the place where the *korale* along with the
swings have been set up. At this new site all the accepters of
pig meat and the novice's close relatives (on the mother's side)
construct *korale*. Land is cleared and two rows of *korale* are set
up facing each other, on the north–south axis. This camp-site is
always slightly in the interior of the forest and not close to the
creek or the coast.

Duration of Tanageru, *Eating, and Feeding*

The division of *eletorakebey* (those who eat) is observable in
terms of the residential pattern, but the coming together of the

[14] With this move in residence and setting up of a new camp-site the whole of
Ongee society divides up in the two groups. Those who are involved in organizing,
coordinating, and conducting the ritual are regarded to be those who conduct the
ritual. The community of Ongees refers to them as 'those who feed!' Those who
feed are the ones who refused the fatty parts of pig meat when the invitations were
being extended. Those who accept the invitation along with the fatty part of the
meat are distinct from the feeders of the ritual and are referred to as 'those who
eat!' (*eletorakebey*).

two divisions is essential for and evident when the Ongees perform the *tanageru*. Those who feed others include the initiate's maternal relatives, known as *beyagee*, and his paternal relatives, *gaakoulotee*. *Beyagee* and *gaakoulottee* do not eat any part of the pig hunted by the novice. *Gayehleg?e*,people related to the novice by marriage, form the core group that consumes the pig hunted by the novice. The *tanageru* ritual involves a sequence of hunting pigs over a prolonged duration of time. It is in the course of this hunting of pigs that the group identified as the feeders, including the novice to be initiated, goes to the forest and hunts pigs. It is both the meat and fat of the pig that is fed to the people who are identified as 'those who eat'. Those who feed them refrain from eating the most preferred part of the pig, that is, the fatty part. The group of feeders is also prohibited from eating the meat of the pig during the first two phases of the *tanageru* ritual. In relation to the groups identified as the feeders and the eaters, the individual to be initiated is seen as sharing the position of both, which makes him distinctive in the series of pig hunts and in the distribution of the excess food which becomes available during the entire ritual phase.

The boy who goes through the *tanageru* does not consume the pig's meat or the fat. However, he alone consumes the blood of the pig he has hunted. It is important to note that the Ongees never consume the blood of any hunted animal, except within the context of the *tanageru*, and even then only the individual whose body is being transformed through the ritual consumes the blood. Pig's blood is also seen as being a liquid food, *gejobe* (a food which is sucked), and is therefore identified with the prime food of children, that is, mother's milk, *neye?neye?*. Its consumption needs no mastication, which creates an identity between a child, a spirit, and an initiate.

While the feeders are responsible for going out to hunt, the eaters are responsible for making sure that the excess pigs which are killed during the *tanageru* ceremony do not get wasted and get consumed by the dogs. In fact, the eaters as a group are responsible for cooking as well as stacking up all the bones and the rotten meat, *nangucumemy?e*, which is packed in a basket and hung on a tree. This creation of a rotting smell and its retention within the camp-site ensures, from the Ongee point of view, that the spirits know of and can smell the large quantity of pig meat which the Ongees have procured by hunting many pigs. This is a conscious projection by men of their having committed offences in relation to the spirits' desires. So much of the meat accumulates in this fashion that generally the entire camp starts stinking of the decaying, cooked pig meat, and the tree on which it is tied becomes the favoured place for the hungry hunting dogs.

Just as the feeders of the pig meat go out day after day, the eaters of the pig meat also cook food for the feeders. This includes only the roasted seeds of jack-fruit collected during the Torale season. In addition to the roasted jackfruit seeds, fresh coconuts and pandanus fruits are also consumed. All these proscriptions and prescriptions pertaining to eating and feeding are referred to as *gilemame*. The food consumed by those who eat the pig fat as well as those who eat the meat is considered to be wet fatty food and is referred to as *obabalade*. The food consumed by the hunters of the pigs is thought to be dry and not fatty, and is called *konkata*. In relation to the *obabalade* and *konkata* foods, the novice's diet throughout the ritual is a progression from *gejobe* (sucked nurturance) to *konkata* (dry non-fatty food). The novice maintains his body in a state of *gy?ole*, light weight. The novice's diet is also referred to as *gy?oleelakuley*, light eating, which consists of pig's blood (wet non-fatty *gejobe*) and jackfruit seeds (dry non-fatty *konkata* food). Blood and jack fruit seeds are collectively referred to as *ekwaikata* (a food which only a person to be initiated consumes).

So important is the aspect of feeding and eating that for the *tanageru* ritual a special kitchen is set up and is referred to as the *andangale*, meaning where the eater and the feeder come together. In day-to-day situations the cooking area is referred to as *kalakale*, which is a combination of the word *ka*, meaning 'where' and *ale* meaning 'children', implying that the cooking area is the place where the question arises, 'Where are the children?'[15]

The distinction between those who are prescribed to feed

[15] Ongees believe that at the place of cooking everybody is a child needing food. In relation to the term *kalakale* it is significant to note that the thatched shelter made by the Ongees is called *korale*, which means a place where the question arises of whose (*ko*) children (*ale*).

and those who are proscribed from eating, *vis-à-vis* the hunting and eating of pigs, divides the Ongees in relation to those who hunt the pig only to feed it to others (but do not eat it). This makes the *tanageru* a ritual occasion founded upon the principles of *gae?bebe* (giving and taking) and *gagoeroyebe* (designator and designatee).

As the father of a boy to be initiated during the *tanageru* ritual told me:

> Many moons back we were many Ongees, today we are few—to do the offensive work—and to bring the offensive work to an end. It is important that all those who can come together should and we all should be tied up. We are to be like the spirits—spirits with whom we give something and take something makes us do a special work which is supposed to be done in a particular place in a particular season. The giving and taking with the spirits is to be the essential thing within the Ongees on the occasion of *tanageru*, and this is *gae?bebe*, since it is important and makes us do the *tanageru* we have to know what each individual and his *korale* is going to work forming the *tanageru* good. Good *tanageru* depends on knowing who tells what work is to be done and each *korale* does accordingly. This is called *gagoeroyebe*.

Thus, during the *tanageru* ritual, the hunters of the pig are proscribed from eating the pig and are supposed to feed the others with what they hunt. In this process they become the *gateyebe*, literally those who make meaning (*gateebe*), that is, designators. The *gateyebe*, who feed others make the eaters of pigs the designated ones with the specific aspect of processing the body of the individual who hunts the pig through the *tanageru* ritual. Those who are thus designated are called *okeye?te*.

Gae?bebe and *gagoeroyebe* are important for the Ongees because they are the two principles which make the *tanageru* happen. For the Ongees participating in the *tanageru* ceremony, the prime function is to come together and process the body of the individual who will go up to the *korale* of Tenyabogalange and enter into negotiations with the spirit of Dare. The Ongees see this as the way in which things can be done as they have always been done. It is essential to send at least one individual up to the *korale* of Tenyabogalange and Dare, so that the 'man-made seasons' may end and the 'spirit-given seasons' may begin.[16] This brings us to the present perception of the Ongee community about the performance of *tanageru*.

The Ongee community, organized on the principles of *gae?bebe* and *gagoeroyebe*, comes together to perform the ritual of *tanageru* as it was done by the ancestors. However, the Ongees see their present way of doing *tanageru*, which follows the method outlined in the myth (refer to Myth no.6), as a means of altering the state of the individual's body so that the ritualized individual can go up to the residence of the spirits. To this purpose the Ongees do replicate the basic procedure of feeding blood to the ritualized individual and making his body light.[17]

The Purpose and Function of Tanageru

For the Ongees *tanageru* is a special form of movement, a translocation that the community enables the individual to undertake. To achieve this 'movement', which has an impact on the movement of the winds, special body conditions have to be created. This is essential since every man is not like the specialized individual, the *torale*, who can travel to the home of the spirits (Tenyabogalange) and be in communication with them. The ordinary individual, unlike the *torale*, has to be ritualized, whereby his body undergoes a requisite change of the kind outlined in the myth about the first *tanageru*, which makes it possible for the individual to travel to the residence of the spirits. For the *torale* such a movement is undertaken at his own choice and mechanism, and it may continue throughout his lifetime. However, each male in Ongee society has to go through *tanageru*, and this makes each male of the community a person who shares his identity with other men in terms of going to the spirits' residence and returning after communicating with the spirits. For all men this is essential, not only because it forms the basis of the shared identity as a person who, like the *torale*, has journeyed once to the spirits with the

[16] None of the Ongees see the performance of *tanageru* as replication of the activity accounted in the myth pertaining to the first *tanageru* (ref. to Myth no. 6), but what they do relate to is that the young male has to be sent up to the residence of the spirits in the same way as the Ongee ancestors did it for the first time
[17] See Myth no. 6, unit E.

help of the society, but also establishes what that individual does for society. Changing the winds and seasons and negotiating with the spirits are thus just aspects of this connection between society and the individual. Society helps the individual to undertake movement and the individual's movement helps the movement of the spirits, which is experienced in terms of winds and seasons. When the people come together to send an individual to the home of the spirits, the Ongees see the transformation of the married young male into a young boy and then into an infant who can go up, just as in the myth of the first *tanageru*.[18] When the initiate returns, he is regarded as a person who has to be re-socialized. The process of re-socialization, which continues after the individual (novice) returns from the home of spirits, is a demonstration of how the individual gets incorporated into the human community. The novice shows that he now is capable of functioning as a full and responsible individual member of the Ongee community. It is this aspect that makes the *tanageru* ritual appropriate for consideration as a typical case of separation from human society, a transition to the spirits, and an incorporation of the individual back into the human community, a total ritual phase of the initiation ceremony as outlined by A.Van-Gennep (1960).

As the *naratakwange* (the initiate) returns from the *korale* of Tenyabogalange, the initiate becomes a person who has demonstrated all the qualities and skills of an accomplished hunter. It is only at the end of the *tanageru* ritual that the individual has the right to hunt alone in the forest as well as at sea. On successful completion of the initiation, the initiate also has the right to initiate other individuals in the future. This is important since the individual learns, in the process of initiation, that he has rightfully become entitled to transmit the knowledge and techniques pertaining to the movements of the spirits, winds, seasons, and things hunted. In the process of undergoing the initiation, the initiate learns the basic logic of *malabuka* and *talabuka*, which makes him much like the 'powerful' *torale*, yet different from the *torale* in a way. The *torale* undertakes the movement without the help of the community at large, though his movements and negotiations with the spirits do aid the

[18] Refer to Myth no. 6, unit K.

entire community. The initiate's trip to the spirits, on the other hand, is wholly conducted and managed by the community (the principle of *gagoeroyebe*) and benefits the community (the principle of *gae?bebe*) in terms of a change of season. Like the myth referring to the first *tanageru*, the present-day ceremony uses the myth and the procedure outlined in it as a map and technique of movement. The 'mythical event' still remains a synchronic cartographic and translocationary principle for the ritual of *tanageru*.

The first myth pertaining to the *tanageru* and the present-day ritual, with all that is uttered, all the gestures that are performed, and all the objects that are manipulated, form a metonymical function for the ritual instead of metaphorical role play. Thus, the myth 'accompanies' the ritual (Levi-Strauss, 1981: 671) and, in the interpretation of the ritual, I intend to show that the utterances, gestures, and manipulations are all basically the same in the context of ritual, myth, and day-to-day life, and are activities for attaining similar goals and objectives, the difference being that the relations between men and the spirits change. Given the relation between men and spirits, and the day-to-day awareness of the accompanying myth, *tanageru* truly represents a kind of ritual which is best described in the words of N. Munn

> as a generalized medium of social interaction in which the vehicles for constructing messages are iconic symbols (acts, words, or things) that convert the load of significance or complex sociocultural meanings embedded in and generated by the ongoing processes of social existence into a communication currency . . . shared sociocultural meanings constitute the utilities that are symbolically transacted through the medium of ritual action . . . by this means a complex world view or aspects of it can, in effect, be circulated within the interplay of the immediate relationships and processes of symbolic (ritual) action [1973: 580].

For the Ongees the relationship between the spirits and human beings is characterized by their affecting each other and having an impact on each other. This inter-effectiveness between man and spirit is a relationship of *enakyu?la* (power)[19]

---

[19] *Enakyu?la*, literally in Ongee language, is used to describe a man pushing a canoe from the shore into water. It glosses aspects of force, strength, push, pull, whereby something is displaced, replaced, or changed in shape.

which makes possible the change between *chongojebe* and *gen-geyebe* within the space shared by the Ongees and the *tomya*.

However, the Ongees articulate their existence in space along with the spirits through their logical models of *malabuka* and *talabuka*, whereby the game of *gukwelonone* (hide and seek) continues. Given this formulation the question arises, what gets symbolically transacted, through what 'medium', and what is the 'ongoing process' in the context of the Ongee community? In the Ongee community the *torale*'s visit to the spirits, the initiate's visit to the spirits, the offending of the spirits to create a temporal phase (*gengeyebe*), and then offending the spirits again so that *gengeyebe* ends and *chongojebe* starts, are all situations that present us with symbolic transaction. They are all aspects of an 'ongoing process' in which men and spirits, through the relation of act and impact, and moves and movements, affect *enakyu?la* (power). The ritual of *tanageru* is a situation which also involves *enakyu?la*.

By focusing on this issue I want to draw attention to the fact that the ongoing process of coexisting with the spirits, living with the acts of hunting and gathering, and not getting hunted and gathered by the spirits, are all part of a 'complex world-view', *malabuka* and *talabuka* being the significant principles in it. In the ritual of *tanageru* and also in a 'non-ritual' context, the 'medium' that is brought into use is the smell carried by the winds, which forms the basis of a 'self-reciprocating asymmetrical relationship' where the men are transmitters of smell and the spirits receive it by means of the winds. Given this symbolic transaction and this medium, can we see the issue of the power of ritual where the body that transmits smell is acted upon and affected? The body is the medium of symbolic transaction.

A child comes to the community of the Ongees when a spirit is trapped in a food substance and in due course comes to reside in an Ongee woman's womb, and is born as a human child. The body of the child (like the spirit) grows up depending on liquid food. The child, on growing up, has to go through an initiation ritual where once again, by feeding on liquid food (pig's blood) the body condition is altered. Once the body condition is altered, the individual (initiate) is believed to be light (*goke?*) enough to undertake a trip to the spirits. Like the

powerful *torale* the initiate returns from the spirits' residence.
The movement of the novice within the *tanageru* ritual context
is not different from the day-to-day context of the spirits'
movements and the *torale*'s movement. Indeed, in both the
ritual and the non-ritual contexts, the movements of human
and spirit bodies display the 'symbolic transaction' of *enakyu?la*
(power) between man and spirits through the medium of individual bodies.

The significance of the body in the course of the ritual and
the non-ritual context is explained by the statements made by
the initiate, the initiator, and the initiate's mother before the
tanageru started. The initiate talked to me about the forthcoming ceremony of *tanageru* and explained what the ritual was all
about, from his point of view,

> I have to be married, then only I can be sent to the home of spirits
> by all other Ongees. Wife and an initiator is important because they
> help in making me go to the *korale* of the spirits. It is with help from
> them and care for me that I can show to all the Ongees that I am a
> *gandema* (brave) hunting Ongee man, worth being between Ongee
> and spirits. I am sent up by all but it is only my wife and her group
> along with my initiator who helps me to come back and be with
> men and women and not end up with spirits forever. If I do not
> come back then the winds and spirits will never come—I will be so
> afraid that I will lose all my teeth in front of them before I could tell
> what everyone has been doing with all that honey and then my
> body if it is all well made through the ritual, specially by my wife,
> her relative and my initiator can come back. It is not just my body
> being made into a small child but also comes back light and as an
> infant. In my going up and coming back I can forget all about our
> place, hunting, working, and living like a human—which I have
> learned while growing up with Ongees. So on returning back I have
> to learn and show to all that I am again an Ongee, a hunter and not
> a dead person, or a spirit, or an infant.

One of the initiators explained the meaning of the term
tanageru as changing the colour *tana* (blue) into *ougeru* (term for
red colours). Though the terms *tana* and *ougeru* are used in a
day-to-day context for glossing blue and red, they have a range
of meanings and each meaning has a specific value within the
Ongee world-view. For instance, *tana* is not just any blue but
blue 'like ash' and *ougeru* is not just red but red as in hot wood.

Tana (ash blue) is not only a description of colour but is also the description for a state of body, especially that of the spirits. Just as the ash can get scattered and cover up everything, the surfaces of a spirit does not have a distinct shape yet can be experienced everywhere. Ash is a greater irritant of the eye than smoke when blown around, and ash causes temporary blindness, which is the main characteristic of the spirits. The lightness of ash is highly respondent to the winds and makes a person blind, so that blue is the colour of the spirits and denotes aspects of the spirits. Ash blue is also associated by the Ongees with the night, a duration when everyone is scared of the spirits, and it is thus important to be in a *korale* with others when it is dark. *Ougeru*, which means hot red, is associated with the red clay paint which consequently makes it a positive value, well within the human domain, as opposed to the ash blue associated with the spirits. Since hot red clay paint causes sweating of the body, it makes a person display his bravery by letting the body smell disperse. It also forces the Ongee hunter to deal with coldness and shivering, which makes the body light and hence conducive to being taken away by the spirits. Thus red is seen to be a colour which protects the human body and makes it capable of processes associated with multiplication and reproduction.

Given the range of ideas associated with *tana* (ash blue) and *ougeru* (hot red), Berogegie, one of the initiators for the *tanageru* ceremony, said:

> We do this ritual so that spirits and winds which had left due to our work can come back. For this purpose the young man has to be made into *tana*, light and spirit-like. Once he is light he is *naratak-wange*, nautilus shell, which can float and go up to the sky from the place where sea and sky meet. Once *naratakwange* comes back he is to be made *ougeru*, like all the other Ongees, different from spirits and children. We do the making of the body (*tamale*) but it is the light body, *tana* of the *naratakwange*, which has to come back and become *ougeru*, heavy and human. In becoming *ougeru* from *tana*, the *gikonetorroka* of *tanageru* makes the Ongee man capable of keeping himself, his relatives, his people, and also the spirits. All that we have taught to each Ongee after he comes from the spirits [after birth] may be forgotten after his second coming back from the spirits' home. On returning back for the second time the Ongee boy

has been brought back a second time, and he has to show to the others that he is brave and like all other Ongees.

It was talking with Choiboi, the mother of the individual to be initiated by Berogegi, which revealed why all the activity pertaining to the *tanageru* ceremony was a serious and solemn affair, making it distinct from as well as similar to day-to-day activity.

It is a *igagame* which makes all the women among the Ongees cry—it is a danger but no *malabuka* or *talabuka* can avoid this 'hide-and-seek' [*gukwelonone*]. My son has to be sent to the spirits, he may never come back, spirits may keep him forever and never shall I see my son again. But this has to be done, every *monatandunamey*, one or many mothers, have to agree about the son going to spirits' home. In sending one child will he return is the danger but he will because we are sending him up—we have stopped the winds and made the spirits angry. Only in the coming of the son will all the women become mothers—no winds for a long time causes a lot of discomfort and hunger. My son has to come back—in his coming back will the season change and so will more children come. I know Berogegi, who is the *muteejeye* for my son's *tanageru*, is good at the work of constructing the ritual. He can send my son up to the home of spirits and bring him down too. He knows it—I know it and my son knows it too since Berogegi was also *mutarandee* to my son. It is always the *mutarandee* who helps in bringing the spirit down as a child and then he alone can become the *muteejeye* to send up the child. Just like the good and the bad spirits—who come giving protection and children and the bad one can come and take away children—death and pain. *Muteejeye* and *mutarandee* are the same—they two form the spirit within the human community, to become a *muteejeye* and *mutarandee* is an honour and skill, as of the *torale*—and to be able to do it every man has to be 'made' through the ritual of *tanageru*. I know all this is essential and it may go without any danger but what happens when my son is away—suppose the spirits realize that we have been tricking them all this season and keep back my son—what we think as the return may never happen, the spirits may never come and it is danger of mine for which all the Ongee mothers will cry—no children—no Ongee no marriage no children — everyone would die— all will become *tomya* and no Ongee would be left.

Putting together statements of the initiate, the initiator, and initiate's mother in the light of the myth of the first *tanageru*, we can formulate what the purpose of *tanageru* is, from the Ongee point of view. This formulation cannot be divorced from the Ongee world-view, which is exhibited in the day-to-day life of hunters and gatherers, a life where man, in order to coexist with spirits, has to manage his movements in relation to the movement of smells.

Hunting and gathering takes place in space in relation to the hunting and gathering of the spirits. The Ongees explain this in terms of *malabuka* and *talabuka*. The presence of danger and the necessity of safety are therefore an ever-present problematic of Ongee existence. This characterizes the conceptions of time and space and the pattern of movements in Ongee life. Since each human body is subject to birth and death, its existence is also connected to the coexistence of spirits and human beings on the island.

For the Ongees a child is a transformed spirit who has come down. Sending the child up, as described in the first myth of *tanageru*, is a process of sending back to the spirits what originally came from them. In the *tanageru* ritual the initiate is turned into an individual who is more like the spirits and has the attributes of *tana*. Indeed, all this happens in a domain that is referred to as *gengeyebe*, where the spirits suffer the impact of Ongee acts. However, the birth of a child, and death of a person in the form of an *enegeteebe*, are normal events in the *chongojebe* domain, where the spirits act and the Ongees suffer the impact. What makes the individual in *gengeyebe* and in *chongojebe* identical is that in both domains the individual body can experience birth and death. Consequently, the body, bodily birth, and bodily death are common elements, events to be found in the two domains that constitute the relationship between the Ongees and the spirits. What differs in the two domains is that in *chongojebe*, the spirits are responsible for birth and death, whereas in *gengeyebe*, the Ongees are responsible for birth and death. Thus, we can conclude that what the spirits do to the Ongees in *chongojebe* is replicated in the *gengeyebe* domain, where the Ongees do the same thing to the spirits. That the spirits become subject to Ongee acts in terms of birth and death shows that humans and spirits interchange their identities and roles. This switching of identity leads to the consequent formation of two distinct domains. But in the

course of doing this, *enakyu?la*, power, is acquired by the Ongees by putting themselves in a dangerous as well as advantageous position. It is this advantage that impells the spirits
to do all the *gugekwene*.

If the Ongees are not in such a situation they cannot acquire
enakyu?la to transform *chongojebe* into *gengeyebe*, to end the
spirit-given season and start the man-made season. It is this
creation of an 'advantageous position' which, from the spirits'
standpoint, is offensive. Moreover, it is offensive because the
Ongees, as in the myth (Myth no. 6, units C and D), break the
basic rule of leaving resources for spirits and not hunting and
gathering the things on which the spirits depend. When the
Ongees collect honey in the season of Torale, they deprive the
spirits of their prime food. Consequently, to have the advantage of consuming 'spirit food honey', the wooden phallus
ritual is performed by the Ongees. This ritual, performed
before the honey collection starts, is actually an enactment of
units H and I from the myth of the first *tanageru*. Even within
the myth, units L and M show how the absence of questioning
about honey by the spirits worked to the advantage of the
Ongees. The association of the honeycomb with the male reproductive organ is explicit in the myth's unit T. However, in
light of the two domains, the two types of seasons, and the
danger in sending an initiate up to the spirits, what becomes
the advantage of the Ongees? When the Ongees consume all
the honey after the ritual of *getankare*, they create their advantage.

When the Ongees gather and feed on all the honey from the
combs in the forest, they ensure pregnancy for the women in
the Ongee community. Once the Ongees have consumed all the
spirit food, the future birth of children is assured. The risk
involved in sending one initiate up to the spirits' residence is
thus offset, and in the end this makes the rituals of *getankare*
and *tanageru* work to the ultimate advantage of the Ongees.

Sending the initiate up to the residence of the spirits is a
form of induced *enegeteebe*, a ritual and offensive effort for a
malabuka (see Myth no. 6, units K, L, M). In their concern over
smell and their use of the *gobolagnane*, the Ongees provide the
basis for the general concern to avoid a *malabuka* because when
a person dies, he or she becomes a spirit. The case of *tanageru*

is different from the day-to-day concern as projected in the use of the *gobolagnane*. If the spirits do not let the initiate return[20] then it implies the death of the initiate, the failure of an initiator, and the loss of a son to his mother. However, performing the *tanageru* is a must because it starts the winds and the spirits start coming to the island again.

This makes the Ongee world, which is dependent on self-reciprocating asymmetrical relations between man and spirit, a truly 'optimistic-alternativistic universe'. If the Ongees succeed in offending the spirits, they get all the honey. This opportunity, an advantage, provides the alternative of potential multiple births, even though the life of the initiate is risked. The risk of the initiate dying, if he is taken away by the spirits, leads to the end of the man-made season and the start of the spirit-given season, a future opportunity for the Ongees and the spirits to coexist. Once the offence committed during the *tanageru* ritual is completed, the winds (which were originally stopped by the ritual of *getankare*) start, and the spirits begin coming to the island. As the spirits visit the island the spirit-given season starts. To get more spirits trapped in the honey-combs, the winds are stopped and the honeycombs are consumed through the ritual of *getankare*. Therefore, the stopping of winds, the expelling of spirits through the ritual of *getankare*, and the starting of winds and ushering in of the spirits through the ritual of *tanageru*, sets the Ongee seasonal cycle into motion. This cyclical motion is attained by the Ongees in terms of birth and death, safety and danger through the rituals of *getankare* and *tanageru*. The two rituals present the exact moment within the seasonal cycle when *enakyu?la*, power, shifts from the powerful spirits to the powerless Ongees (that is, after the effects of *getankare*), and then goes back from the Ongees to the spirits (that is, after ritual of *tanageru*). It is the dynamic aspect of *enakyu?la* which makes it possible to arrange the relation of man and spirit in two distinct systems, which form the domains of *chongojebe* and *gengeyebe*.

The Ongees at Dugong Creek took forty-five days (mid-May to end of June) to complete the *tanageru* ritual. For the sake of analysis, and in accordance with the objectives and goals of

---

[20] Contrary to situation outlined in Myth 6, units M and N.

the ritual, I have divided the ritual into five phases. The phases are distinguished in accordance with the distinctions perceived in the initiate's body state. The prolonged duration of forty-five dry, windless days from May to June is quite arbitrary from the Ongee point of view. The duration of the *tanageru* ritual depends on two things—first, the wind conditions, and second, the rate at which the pigs are available for hunting.

PART THREE

Tanageru

CHAPTER 6

Phase One: Preparations for the *Tanageru* Ritual

After the invitations have been sent out and people have come together for the ritual of *tanageru*, all the camp-mates set up new *korale*. The individuals who are to be initiated are marked as *eneedabatanebe*, implying an isolation. The initiate's isolation involves permission to take his wife to a *korale* that his father makes in the forest. The male 'to be initiated' and his wife are allowed to stay in this *korale*, away from the main camping ground where *korale* for all the others are under construction. The two-night duration for which the initiate and his wife are allowed to live outside the main camp-ground are seen as giving the wife a chance to express her *menyakuttu wanabe*, affectionate crying, which is essential since her tears are believed to be a source of remembrance for her husband that enables him to come back to the Ongee community. The Ongees believe that this form of crying can take place only between husband and wife, and that no one else should know how intense it was. Thus the *korale* built by the initiate's father is covered all over with thatch, which is never done in day-to-day life, the exception being the case of a young girl after her first menstruation.

After two nights all the women go to the forest and bring the initiate and his wife back to the newly set up camping ground. As the young male to be initiated comes to the new camp-ground accompanied by his wife, the formal welcome is marked by a silence. Nobody talks to the initiate, and the wife also keeps quiet. The father and father-in-law come forward with all the new bows and arrows, and take the lead in the procession that includes the initiate and all the women. The father and the father-in-law direct the procession to the centre of a long elongated sleeping platform made for the *tanageru*. The novice is then made to sit on the platform, and the strings of the new bows are plucked by the father and the father-in-law, this producing a vibrating sound. It is with this sound of the new bows that the silence is broken, and at the end of each series of sounds of the bow-strings, the camp-mates in solemn low-pitched voices, repeatedly chant the words '*Durru durru durru!*' *Durru* is not only the onomatopoeic word in the Ongee language for the bowstring's twanging, but is also the word for the noise of the monitor lizard's call to the spirits, to which the spirits respond with lightning and cloudburst (Radcliffe-Brown, 1966: 145,151).

After this, all the *eneyobe* (respected elders) of the camp come to the wife of the novice and in loud clear voices say to her, 'We are ready to send your husband away, so he shall from now on sleep away from you, with all of us on the sleeping platform we have made'. After this declaration, the wife is expected to keep away from the jointed elongated *kame* placed under a single roof. Every alternate day the novice sleeps with all the married women of the camp and then with the married men of the camp. This creates a situation where none of the married couples sleep together throughout the duration of the *tanageru* ritual. When women surround the novice at night, all the men go to sleep in their respective family *korale*s. The situation in which the married men sleep with the novice, away from the family *korale*, is referred to as *ekatowejaloke*. Similarly, the situation when the women sleep with the novice, away from their respective husbands, is referred to as *ekwaawakabe*. For the Ongee *ekatowejaloke* and *ekwaawakabe* are all part of important aspects of isolation, *eneedabatanebe*. Isolation, implying rules for sleeping together as well as separately, is of crucial importance (*see* Radcliffe-Brown, 1964: 157,34) to Ongee metaphysics. Ongees believe that when they are sleeping the 'body internal' goes out from the 'body external' and collects all the smell left behind in the course of the day's movement.

If all Ongees sleep together then we will be safe and then in the morning we can know what all of our *eneteea* [body internal] saw [this helps the initiate to finish his hunting soon], because we all find out what everyone's body internal saw while the body external

was sleeping — sleeping together makes it possible for the body
internal of all of us to travel together!

Given this reasoning, sleeping with the spouse distracts the
body internal, which tends to come back and remember all of
what was seen during the 'dream time'.[1] After this declaration
and the organization of the sleeping arrangement, the *eneyobe*,
the respected elders, are all thanked for *matebe*, giving consent
for the start of the *tanageru* ritual.

The early morning of 22 May 1984 was like any morning.
The whole environment was at a standstill. None of the leaves
were moving, none of the birds were chirping, the dragonflies
stood still on the calm waters of the creek and sea. Two days
had gone by since the elders had been thanked for the consent
to hold the *tanageru* ritual. No significant wind conditions were
to be experienced. This made the heat, at a temperature of 85
to 90 degrees Fahrenheit, feel very intense.

Just as the whole environment grows very still, so does the
camp wait very pensively. For the past two days, the fathers
and the fathers-in-law of the boys to be initiated have been
sitting still, staring at the south-eastern horizon. At the peak of
the afternoon of 22 May 1984, the temperature starts dropping.
Dramatically the light conditions change into partial darkness.
All the forms of life on the island, including the Ongees, become
restless. The men gathered on the shoreline, staring at the
south-eastern horizon, start murmuring '*Durru-durru-durru!*'
The murmur builds up in volume and the darkness sets in.
Everybody except the novice moves to the shoreline in a very
slow and composed manner. The young children are firmly
told by their parents to remain close-by and be quiet. Everybody wants to see what the men were waiting to see on the
south-eastern horizon.

It was Teneyabogalange leaving along a path known as
tegule (water spout). For the Ongees *tegule* and Teneyabogalange are the final forms in which all the spirits and winds go
up. Tegule, a pathway connecting the sea and the sky, is a
funnel-shaped water spout. The thundering noise ('*Durru*'), the

---

[1] Some of the Ongees also told me that during the duration of the ritual husbands
and wives sleeping separately helps the Ongees refrain from acts of *ketukabe*,
'comforting' (copulation).

darkness, and the swirling winds above the surface of the sea create a huge funnel. The funnel's girth increases as it rises and goes away from the land, swaying and whirling. It is a sight that the Ongees perceive and relate to with an aura of fear, and is indexical of their actions. It is the funnel formed on the south-eastern horizon, on which the spirit Dare (after becoming very angry, referred to as Teneyabogalange), goes away to send down more rains and winds.

As the Teneyabogalange goes away the residents of the camp come back from the coastal area and collect on the central camp-ground. Now the sequence of hunting pigs for the *tanageru* ritual is to start. The novice's mother's brother brings two sticks from the trees that have fallen down on the coast during the recent cyclonic conditions. The sticks are measured against the height of the individuals who are to be initiated. Once the first stick has been cut to the height of an individual, it is given to the father of the novice. The novice's father cuts another stick of the same length and gives it to the novice's mother. The novice's mother and wife then take three forked stalks and plant them firmly in the ground. Then the two sticks cut by the novice's mother's brother and the novice's father are tied down horizontally on the forked branches, forming a scaffold on which the skulls of the pigs hunted will be hung. The direction in which this constructed scaffold faces is always south-west, the direction from which the next season will be coming.

The height of the novice[2] who is going to be initiated is converted into the length of the stick. The height of the novice, which is the same as the length of the stick, is a measure to fix the total duration of the pig hunting for *tanageru*. Throughout the duration of the *tanageru* no Ongee could tell me when the ritual would be over, but they would always point at the scaffold and say: See that (pointing to the scaffold); it is

the body of the *naratakwange* (novice), every time he feeds on the

---

[2] In the course of my field-work I randomly selected twenty adult men and women and measured their average height. Average height for men was four feet, four inches and average height of women was four feet two inches. The tallest male height recorded was four feet six inches and of females it was four feet five inches. Minimum height recorded for men was four feet two inches and for females was four feet.

pig's blood the pig skull will be hung there and when all the length of the scaffold is covered with the skulls it will end the *tanageru's* pig hunting.

For all the Ongees, this particular scaffold is the externalized projection of the novice's body conditions. According to Teemai, one of the initiators (also father-in-law of the novice):

As the novice becomes lighter and lighter by his feeding of blood and hunting pigs the sticks will become heavier and heavier with the pig skulls, when the sticks are half covered we know that the novice is ready to go up and come back from the spirit's residence. We will keep all the skulls bound together with the woven cane. His [initiate's] wife will save it in her *korale* for all others to see till her husband becomes a *mutarandee*. Then only those skulls can be buried in the forest, and all of those be thrown into the sea. So that more pigs and dugongs are born for being hunted by the one to whom the one to be initiated is the initiator.

The scaffold constructed in accordance with the initiate's height, for the purpose of hanging pig skulls, is called as *naratakwange-ye gotterange tulukuceye*, literally meaning 'nautilus shell form of the novice that gets pig skull for keeping'. Once the scaffold's construction is completed the *muteejeye* (initiator) goes to the novice and recites:

Yuva etheebencenenegee??-
You are now dying
Yuva gaikabanka utokwobe naratakwange??—
you will be coming out [birth] as a nautilus shell
Yuve ethhe eolobe-naratakwange Ebobe maa naratakwange??—
you now will move as nautilus shell
    Fear not nautilus shell
Inkebe ukotatanka ceragee gaeebatee Enekutata- kivejangnene??
going upward go quickly briskly We shall all wait for you
Me ethee yuva mutarandee wa mutejeeye
I am now your Mutarandee as well Muteejeye
Muteejeye wa yuva ule ule gawakobe??
(Thereby) I *Muteejeye* will again and again give you *gawakobe*
    (form of discourse)
manota ateelabe gagee enekutate ateelebe maa??
with me [alone] you talk. All right! With everybody talking is no
    good for you.
maye?kwe manota inkebe tomya-ye korale- lolobobe ma??

Walk behind, with me onward to the home of the spirits —
 shiver not.
Gayabatee ceragee, koyrabe Yuva ebobe maa, gobokuebe ma??
Briskly go floating You fear not, forbidden is not to
ateelebe-lemolake?e gawakobe tanageru enekutate
talk-however be silent *gawakobe* (discourse of *tanageru*) is
onkoboykwa tomya dare Umuka maa??
for only all the spirits Sleep not
ngijayka kuttu kuttu tolakebe olo wa geduba
From you many many damages of lightness and heaviness
 will come about
Yuva ale wa tomya wa naratakwange Ethee ceragee Tanageru.
You are the child and the spirit and the nautilus shell. Now
 will go on the *Tanageru.*

The declaration is made by the *muteejeye* to the *narata-kwange* (the one to be initiated). This recitation is done in a very stylized manner and is made possible by the distinctive use of the extended glottal sound at the end of the sentence (represented here by ??). Each sentence is thus uttered at a somewhat slow pace. In other situations in which the Ongees recite or sing, the repetition of lines is practised, but this does not occur here. All this makes the initiator's declaration to the initiate a kind of talk that the Ongees distinguish as *gawakobe*. *Gawakobe* as a distinct form of talk is related to selected conversational contexts and speech events involving the spirits and an Ongee individual, ancestor and his or her child, or a child and the *mutarandee*. In the Ongee language, talk is *aateelebey*, but talking with the spirits, an initiate, or a child is not always *aateelebey*. The communications between an Ongee and a spirit, a child, and ancestor, and an initiate and an initiator are *gawakobe*. What makes all three cases *gawakobe* goes beyond the identity shared by spirit, child, and initiate. The prefix *gawa* in the word *gawa-kobe* is the root for the verb to look in the Ongee language. It is this prefix that fixes the special position of *gawakobe* distinct from *aateelabe*. According to the Ongees, all the conversations held between an adult and a child, an initiate and an initiator, and an Ongee and a spirit are such that 'Telling is not only a way of seeing but also a way of showing!'
 For the Ongees the distinction between seeing and showing constitutes aspects of training, socialization, and learning im-

parted by the one involved in the discourse of *gawakobe* (i.e. the adult, the initiator, and the spirit). Thus the adult in relation to the child, the initiator in relation to the initiate, and a spirit in relation to an Ongee are all *eneyatelabe*, the ones who tell about seeing, which has to be learnt by those who listen to the *gawakobe*. Once the listener has heard about seeing, what is acquired and retained is *eye?e*,[3] knowledge which makes it possible for the hearer to show what he has seen.

In a simple, day-to-day context, a child accompanying his father on a hunting expedition demonstrates the functions of *gawakobe*. The son accompanying his father on a hunt is not allowed to talk, he has to maintain silence (*lemolakeye?e*). No talking is permitted (*aateelebey* is proscribed). However, the father tells the child all about the forest, gives him clues about where the animals are and how to track them, and how to avoid being traced by the 'hunting spirits'. This (what the father says, in terms of seeing) is all *eneyatelabe*. The son has to learn all this so that he too can go to hunt on his own and 'show' what he saw and that he learned as a 'way of seeing'. This is the *eye?e*, knowledge (a way of seeing). Practice based on this knowledge makes possible the way of showing. The aspect of seeing and then showing can also be traced in the case of the spirit communicator, the *torale*. The spirits communicate to the *torale* 'what to see' and 'where to see'. Upon returning from the spirits' residence, the *torale* 'shows' the other Ongees what he had seen while he was 'above the forest'. In the course of the *tanageru* ritual the initiator makes it possible for the initiate to see and show his accomplishments that involve going to the residence of spirits and then returning.

After the initiator delivers the discourse of *gawakobe* to the novice, the novice is subjected to various rules. The basic rules that characterize the behaviour of the initiate are as follows:

1.   The novice has to maintain *lemolakeye?e*, silence. No one except the initiator talks to the novice, and even he must talk in a whisper. For the novice to interact verbally is completely proscribed.

[3] The literal gloss for *eye?e* is 'something known to a person which cannot be taken away but could be given'. For example, honeycombs in the forest and the ability to make clay pots.

2.   The novice has to keep his eyes partially shut, and in all his movements from place to place, has to follow the initiator.

3.   His food intake is limited to only what the initiator gives him and he may eat only when food is given to him.

4.   The novice is to hold his stomach with both hands crossed. This is essential because his body is undergoing changes and is thought to be off balance, which may result in his falling down while moving about. This style of walking with the hands folded and crossed at the waist is called *gedentube*.

5.   On the instructions of the initiator the novice goes out to hunt pigs, and this series of hunting expeditions is called *eneratetangeyabe*. The most important consideration is the place where the pigs are to be hunted by the novice. During this hunting, the actions and gestures outlined by the above-mentioned rules (nos.1 to 4) have to be complied with.

6.   The path on which the group, along with the initiate, leaves on an *eneratetangeyabe* has to be avoided when returning to the camp after hunting the pig. The main idea is that the novice, after a successful hunt, should avoid being seen by the women.

Next morning (23 May 1984) the camp got up unusually early, just before daybreak, and everyone made arrangements to go out to hunt with the initiate. The initiate now behaved in accordance with the rules outlined by the initiator in the discourse of *gawakobe*. The group leaving for the hunt had to include the initiator and the novice's father and father-in-law. The people accompanying the initiate carry all the weapons, the fire, and the basket. The accompanying people also arrange for a few dogs to accompany the hunting party. Every family at the camp contributes at least a dog for the hunting expedition. Canoes are placed at all the points where the group may have to cross water. (The arrangements pertaining to the canoes are all made on the night preceding the day when the group leaves for the hunt.)

At the Forest: The Initiate, the Initiator, and the Pigs

The expedition of the initiate, accompanied by the initiator, father, and father-in-law, has significance at two levels. The first is that prior to this occasion the novice has accompanied others to hunt pigs, but has never gone alone and hunted pigs or turtles. In the past the novice has accompanied the hunters on hunting expeditions, but he has never been allowed to commit the final act that actually takes the life of the pig. Thus, he assists in hunting but not in actual killing. All the non-initiated males are proscribed from cutting pigs, an act which releases blood. As the Ongees put it, 'only spirits, women, and initiated men can be involved in cutting something that would cause release of liquids like blood' (myth of *tanageru*, unit B).

The second point of significance is that the novice who hunts at the time of his *tanageru* is allowed to hunt only *tamalelebe*, full-grown male pigs. The full-grown pig is to be hunted in a particular way called *gayekwabe*. In this method of hunting, the dogs stalk the pig but are not allowed to attack it. The pig is killed by the novice with new arrows made for this particular occasion. Once the pig's body has been impaled by an arrow, the novice has to walk up close to the pig and slit open its stomach with a knife. What is emphasized in *gayekwabe* is that the dogs are not afraid to corner the pig, and the hunter is brave enough to go close to the pig and is strong enough to cut open its stomach. In the course of this action the novice practically has a hand-to-hand fight with the pig and restrains the dogs from attacking it after the arrow has been shot.

In the ordinary context, the hunting of pigs is very different. Generally, before shooting an arrow, the hunter locates himself near the spot where the dogs, *uweme*, have cornered the pig. He also positions himself close to a tree, so that as soon as the arrow has been shot he can climb the tree and avoid being charged by the 'angry' pig. However, during the *tanageru* hunt, the novice is not allowed to climb a tree after shooting the pig. This climbing of the tree is called *beletee-kwotobe*, and the Ongees believe that by doing this 'not only do we get away from the pig's path of "coincidence", but like the spirit we get up so that the presence and the absence of the smell confuses the pigs as well as the spirits'.[4]

In the day-to-day context the *uweme* (dogs) generally succeed in practically killing the pig by cornering and charging at it. In such a case the pig is thought to be a weak and cowardly animal. Pigs who are weak and cowardly when they are hunted are called *gababe*. Sometimes the hunters reach the spot where the dogs have cornered the pig by listening to their barking. Upon reaching the spot, the pig is killed with the help of arrows and/or knives, and such hunts are called *gebekyabe*.

As the group with the novice reaches the forest they start paying attention to the direction in which the majority of the dogs are going. Some accompanying people stay at the point from where dogs have started on a trail. The novice, along with the initiator and the father and father-in-law, rapidly follows the excited pack of dogs. This is not easy because the initiator in front, who holds the bows and arrows, has to lead the way by cutting through the thick undergrowth, and he has to be quick so that the dogs cannot scare the pig. This procedure may be of a short duration or may be long, and may end up in accomplishing nothing. A pig that is not male and full grown is not counted among the number of pigs killed for the ritual. When the initiator sights the pack of dogs who have cornered a pig, he hands over the weapons to the novice and stands back to watch the novice kill the pig in accordance with the *gayekwabe* procedure.

In a short while the barking of the dogs ends abruptly, merging with the resonance of the large wild boar groaning loudly, and the last fatal 'cut' is made by the novice. All the other men, who have stayed back at different points in the forest while the novice approached the pig in silence, are now aware of the bloody scene. This scene is essential for the ritual. The novice's father, who holds the fire, along with the initiator, approaches the location (generally 20 to 40 feet away) where the novice is waiting for them with the dying boar. The novice waits for the initiator and his father so that the actions taken by him may be endorsed and approved.

[4] Interestingly, Ongees while hunting turtles jump out of the canoe after implanting the harpoon. This is referred to as *geeteekwotebe* and is also explained in the same manner as spirit and pig within the context of *beletee-kwotobe*.

The initiator turns the pig, still struggling for its life, on his back. He takes his machete, heats it over the flame carried by the novice's father, and gives it to the novice. The initiator holds the pig's hind legs, the novice puts his right foot on the boar's neck, and facing his initiator, with a single stroke cuts the boar's phallus and genitalia. At the time of this slitting (which has to be accomplished in one cut) everyone shuts his eyes. This restriction on seeing is called *genge-geje?jele*. Nobody except the novice should see the actual cutting operation, which is referred to as *gananbube*. The restriction on seeing *gananbube* is intended to make sure that no one but the novice remembers where he made the 'offensive blood-letting cut' (Myth no. 6, unit E, K).

Gananbube is performed very solemnly by the novice. Complete silence is maintained and after the 'cutting' everyone releases a long and heavy sigh. The novice is directed by his initiator to take a few steps away from the bleeding pig, and fire is placed close to him. The father now starts collecting the blood from the pig in a quickly assembled bundle of leaves folded and tied together with rattan.[5] The father, the father-in-law, the initiator, and the novice, with the pig placed on the back of the father-in-law, come to the spot where the rest of the group is waiting. As the group gets together the initiator declares:

> *Ototukwene* (collection of blood is over), we have killed the pig and we did not scatter any blood for the spirits!

Once the whole group of hunters gets together, it starts walking back to the camp. The father and the father-in-law take the responsibility of carrying the weapons and the pigs. The novice walks slowly, holding his stomach, and supported from the back by the initiator. This particular posture is called *gakute-totamale*. According to Tai, one of the initiators, the *gakute-totamale* way of walking is essential, because

> the novice is not comfortable since he has not eaten anything. He has to have *mandaneyele* (fast) so that his body becomes light. Since his body is changing he may fall down so he has to hold himself at

[5] In everyday context the blood released by killing and cutting the pig is all thrown upwards in the forest. This is done for the spirits in the forest who are out hunting pigs and humans, and is called *galujebe*.

the waist. The novice is also reluctant to go back to the camp where all others are waiting. The novice is ashamed of his bad work, involving hunting of pigs, the spirits are going to be angry and they are going to get him.

While the group of men is out hunting, the women in the camp divide up into smaller working groups to fulfil various responsibilities. *Gotatebogeru* (string ornaments) are given the finishing touches and they are all brought to the novice's mother. Red clay paint is prepared. The *wabe*, the central camp-ground enclosed by the sleeping platforms, is cleaned. Fire-wood is collected and stacked. As the women finish with these responsibilities they start what is referred to as *enna-kwu-tuwelakube*, collectively waiting for the return. While waiting, all the women tie strips of bright yellowish-green pandanus leaves on their foreheads and around their waists. These strips are treated as markers to indicate that the women are im-patient, concerned, and eager for the novice to return.

As the novice and the men return the waiting women stop referring to the novice by his name. The novice is referred to by the women as *gitekwatebe*, 'One who is hunted'. Some of the young girls hide among the shrubs growing at the creek that will be crossed by the men before reaching the camp-site. It is the responsibility of these young girls to look for the returning party. If they spot the men coming back with bundles of leaves, they run back and report to the women in very soft voices, '*Ethee gukukwayabe kuttu kuttu bebele!*' (now is the smell hidden with many many leaves).[6] When this is reported by the girls, the women in the camp know that the hunting party has been successful in getting the right pig.

Though the men started to go out to hunt on 22 May 1984, they did not meet with any success until 26 May 1984. The three days of failure in locating the right pig, and the repeated departures from the camp did not have any effect on the camp mates' attitude, but physical exhaustion was evident. During the three days, the failure was referred to as *nakwangetlakwebe-igagame*. As a day of failure ends the whole camp follows the rule of not singing, since singing attracts the food and that

[6] The leaves of *toijage* of the tree *teigatale* spotted by the girls are the ones which men bring from the forest to tie up a curtain under which the initiate is served pig's blood.

would make the hunting of the pig by the novice too easy. It should not be overlooked that in the course of the three days nothing was simple and easy for the novice, since all the restrictions pertaining to food, walking, and sleeping prevailed for him. Women generally express concern and 'maternal feelings' by visiting the novice's mother and telling her, 'Our child is dying but he will return soon—it has always been like this—see he doesn't want to leave all of us who have given him milk!'

As the night sets in, people settle down to eating. Some eat jackfruit seeds, while others feed upon the pigs brought in by the novice and the hunting party. The pigs brought between 23 May and 26 May were either too young or were female or both. So the initiators and the novice, who refrained from eating, had already started feeding the camp. Some of the reasons for failure in hunting the pig given by the Ongees were as follows:

> Our dogs went out of way, we did not keep them well—they are not hungry enough to run after pigs. The dogs that did trace the pig were afraid and let the pigs go by—see the pigs are in the forest—ready to be hunted for the *tanageru*, however we should take different dogs. It has been too hot and it has now rained in the forest this rain would cover up all the path we have walked on and left our smell on—this would make the pigs forget that for the last few days we have been going to the forest to hunt them. We have been going out to the forest very late—we should start early in the morning and for that we should all go to sleep early—so that we all have more time to dream and know where to succeed in hunting the pigs!

Around five on the evening of 26 May 1984 the hunting party returned successfully. The novice had succeeded in hunting two full-grown male pigs in accordance with the rules of hunting. As the returning party crossed the last creek on the way to the camp-site, the fire taken to the forest was left behind. On getting off the canoe the novice and the initiator stayed together, walking slowly in silence, and the novice held nothing but his own body. The father of the novice takes the lead in the procession and walks ahead. It is the father who is supposed to now inform the waiting women and girls that the novice and the group are entering the camp. It is the responsibility of a camp-mate to inform others about an arrival, especially when something new or a new person enters the camp. This is called

tananey, and during *tanageru* as well as in all other cases, the
buttress of a tree is beaten with a log of wood or with a machete.
This produces a loud resounding sound.[7] In the case of
tanageru, *tananey* is essential because it not only tells the women
that the party is returning after a successful hunt, but above all
tells them that something new is coming. This new element
approaching the residential area is the novice. He is different
from the person who had left the camp. He has now succeeded
in hunting a pig—an offence that puts him a step closer to his
death and the encounter with the spirits. When the women hear
the *tananey* they all shout the word *'eneyayebo?be'*, which means
we are afraid, three times together. This collective fear expressed by the women from the camp (about 300 feet from the
place the *tananey* signal was sent) is called *tuatey*. On hearing
the *tuatey* the novice and his companions start walking towards
the camp. Meanwhile all the women go to the novice's mother
and in very soft and sympathetic voices say

> Our child is dead he is going to go away but he may come back and
> with his return more children will come—for he is now just a
> *gande?be* (something to be afraid of)!

The transmission of the *tananey* and the reception of the
tuatey frame the first sequence in the series of transformations
of the novice. It is from this moment that the whole community
accepts the change in the identity of the novice and then starts
to further transform the novice's actual physical and bodily
existence. The break in the natural noises of the forest caused
by the beating of tree buttresses sets the stage by which the
change of the initiate into something dangerous is acknowledged by the whole community.

The discourse of *gawakobe* had set up the terms for the
ritual. After the *gawakobe*, the hunting of pigs and the specialized acts and gestures had started. Now the activities that

[7] In the early phase of my field-work, going from camp to camp, walking for two
to three days through the forest, the accompanying Ongee would send a message
to the camp where I had not yet been introduced by beating the buttress of a tree,
i.e. *tananey*. As time went by Ongees stopped doing *tananey* when I visited them in
different residential areas. Ongees believe that it is not the sound of the tree that
communicates to them what is approaching but that beating the tree disturbs the
birds and the cicadas, and their silence for a short while and then repeated intervals
of forest noise and absence of it is what tells the Ongees that somebody is coming.

manipulate the novice's body and physically transform him begin.

Arrival of the Novice at the Camp-grounds: Deconstructing the Body and Nautilus Shell

As the women's *tuatey* is heard by the returning hunting party, they start walking towards the camp-ground. On entering the centre of the camp-ground the pigs and the leaves are put down. Everyone comes to examine the pigs. They lift the pig, turn it around, and check the wounds on its body to confirm that the right procedure has been followed. While the men are doing this, the father and the initiator take the novice to the *kame* where all the women have congregated. The novice is left in the company of the women. All the women (except the novice's wife) congregate at the extended sleeping platform and surround the novice. The novice's mother and his paternal and maternal aunts hold him tight while he sits on his mother's or aunt's lap (form of *enegeteebe*). Other women surround the core of women who hold the novice. As everyone takes her respective position, all the women's bodies are in contact with each other, complete in a circuit. Their bodies are interconnected by placing the left hand around the shoulder of the next woman and the right hand on the shoulder of the woman in front.

The formation in which the women sit is called *gattua-narat-wa*, which means a nautilus shell broken in half. This sitting formation derives its name from the way in which the women arrange themselves around the novice. Women around the novice are like the cross-section of a broken nautilus shell. This reasoning was given to me by one of the initiators. However, as the women settle down, the initiator brings a whole nautilus shell and splits it into half by striking it with his machete. The initiator then hangs the split half nautilus shell on the thatch right above the platform where women are sitting with the novice. On asking the initiator about the shell he had split open and hung on the thatch, I was told,

> We have done the bringing of shell and the pig to make the shell go. Now women will do the breaking of the shell [novice] just as I did! [Myth no. 6, units E, K].

What is being represented and communicated for the Ongees through *gattua-naratwa* is that the novice's body transformational process has started. This becomes evident in the
purpose and form of crying that the women start after the
initiator has split open the nautilus shell. *Wanabe*, crying in the
context of the women placed in the formation of *gattua-naratwa*,
is called *enatatanuye*. This is a distinct form of crying in which
the women are regarded to be inhaling all the smell from the
body of the novice and also sucking the body liquids from the
novice's body. Since the women 'are consuming so much from
the body of the child, they are killing[8] the child they have raised
by feeding various liquids'. It is in this explanation of the
enatatanuye given by the Ongees that crying becomes a representation of various sentiments.

The significance of replicating the form of the chambered
nautilus shell for the Ongees goes beyond its reference in the
myth of the first *tanageru*. It is important to consider how the
Ongees describe the shell. The chambered nautilus shell, *naratakwange*, is distinctive for the Ongees because 'it is a big shell',
often used as a container for liquids, but above all it is regarded
to be the only shell that can float. The Ongees pointed to the
chambers inside the shell that were split open and said, 'When
liquids come out of the small sections of the shell the shell
becomes more shell. New shells are born out of the shell whose
liquid is lost. This [loss of liquid] makes the shell capable of
floating!'[9]

---

[8] Women are associated with spirits (Myth no.6, unit A) and are capable of (just
like spirits) giving and taking life. But it is important to note that sucking the juices
is also a way in which spirits kill humans after they have had an embrace.

[9] For Ongees, liquids passing out through the chambers of the nautilus shell cause
destruction of the shell as well as multiplication. The Ongee explanation for this
was:

> Our ancestors have told us that whenever visitors [outsiders] come and give
> us iron—something heavy which we could use to make *maonole* (tools and
> weapons) they would want things from us too. They like to have nautilus
> shells always—especially white man and Burmese wanted it—since they
> give us iron [which is heavy] we would give them shells that were heavy as
> well as dead. But the shells given to outsider were not the ones from which
> water had come out. This made more white men come for shell—many
> Ongees then started dying at the coast, in war. It was our ancestral spirit
> who told us to remove the inside and inside water of shells—and more
> Ongees will remain alive. This we did and since then white men stopped

It is in the description of the chambered nautilus shell that one can see what the women are doing to the novice. The women sitting around the novice forming the *gattua-naratwa* replicate the structure of the nautilus shell, once it is split open. The women crying and 'inhaling the smell' from the novice's body (*enatatanuye*) are doing what happens to the shell, in terms of liquids passing out of its chambers. The outcome of what the women do, in terms of *gattua-naratwa* and *enatatanuye*, is identical with what happens to the nautilus shell. The novice, after going through the *enatatanuye*, becomes light and capable of floating exactly like the nautilus shell.[10] Once the *tanageru* is over, the spirit-given seasons will start and new Ongee children will be born because of the spirits trapped in the fruits and honeycombs. It is this aspect that forms the homology: with the breaking and loss of liquid from a chambered nautilus shell, more shells are born; the women who do *gattua-narratwa* and *enatatanuye* to the novice's body get more children.

Under the Curtain, Over the Pigs, Sitting to Go Up

As the women with the novice continue with *enatatanuye*, the men make the *enerayale*. An *enerayale* is a curtain made out of leaves and rattan. Small bundles of *teigatale* leaves are tied to the rattan and the rattan is raised up to five feet. This forms the *enerayale* that is hung across the camp-ground. Thus the *enerayale* divides the sleeping platform and the camp-ground into two.

Under the *enerayale* the male pigs killed by the novice are placed on their backs, and the legs of the two boars are tied together. By placing logs of wood on the sides of the upturned pigs, the initiators make sure that the pigs are firmly enough

---

coming and we accumulated shells and we could give away the shells when we wanted rather than give it whenever the visitors came and demanded it.

In light of the history of Little Andamans since 1850, Burmese traders use to visit the island and demand nautilus shells. Since the outsiders demanded living shells Ongees had to search for them. (Living shells have a higher market price since they can be processed and the natural gloss can be retained.) However, this searching for and demand for shells led to many confrontations between the Ongees and outsiders (Portman, 1899).

[10] This action of the women corresponds to unit K of the myth of the first *tanageru*

placed for the novice to sit on their stomachs. This 'seat',
formed by tying the pigs together under the leaf curtain, is
called *ene?yabelabeebey-ete-betenalabe*, which means now turning upside down. As all the preparations are completed the
initiator goes to where the women are sitting surrounding the
novice. The initiator lifts up the novice in his arms, as he would
a helpless infant.

The initiators then make the novice sit on the pigs. The
novice sits close to the pigs' groins, which were cut out in the
forest right after the pigs were hunted. The novice sitting on
the pigs, with his body slightly bent forward towards the pig's
mouth, is under the leaf curtain. This posture and location, a
unique moment, is referred to as *ene?yabelabeebey-ete-betenalabe*,
which in the Ongee language means 'now is the turning upside
down'. The novice sitting on the pigs is referred to as *erabota?be*,
which means to go upwards sitting down.[11]

The curtain is made of the leaf of the *teigatale* tree (the leaf
of this tree is referred to as *toijage*) that has its own significance,
because it is the only tree which sheds its leaves after Dare has
left the island. When Dare comes to the island with its accompanying winds, the Ongees believe that the spirit and the winds
of the Dare season are slowed and obstructed in passing by the
teigatale tree's leaves. It is because of this that the leaves of the
teigatale are unique in coloration. One side of the *teigatale* leaf
is deep green and the other is light white. Thus the Ongees see
the tree of *teigatale* as having a special affinity with the spirit of

---

[11] In Ongee language sitting down is glossed as *belakwe*, which means not moving
or staying. The word *eneyamege* is used to convey the idea that someone is sitting
upon something, for example, *yuva eneyamega kameka?* means you sit upon a
sleeping platform. The term *erabota?be*, which means sitting down to go up, is only
and exclusively used for a person sitting in three contexts. These are of a person
dying, women sitting for childbirth, and the novice going through initiation. All
the three contexts invoice aspects of vertical movement (related to spirits) taking
place on horizontal axis (related to humans). The underlying principle in all the
three situations where the individual is completely motionless are referred to as
sitting up to go up (*erabota?be*), because in them are hidden implications of the
possibility of movement on the vertical axis. In the case of women going through
labour pains the child to be born may be dead—in case the spirit coming out of the
womb goes back to the places associated with spirits. Thus, in childbirth the spirit
is made motionless, specifically in relation to the potential movement along the
vertical axis. In the same way, the person dying is seen as going up the tree which
connects various places.

Dare, which 'walks on it', causing a change of coloration on one side. However, the leaves do not fall from the tree even when the intensity of the wind's movement causes all the other trees' leaves to fall.

Since the leaves have been brushed by the spirits and the winds, in the season of *tanageru* they are seen to be potently capable of hiding the smell of the pig hunted by the novice. Consequently, the pig hunted by the novice during the *tanageru* is also covered in leaves of *teigatale* when the group returns from the forest. It is because of the correlation of the *teigatale* as a botanical category with meteorological changes that the curtain made from *teigatale* leaves become significant for the initiation ritual. The location of the *enerayale* makes this clear.

The rows of sleeping platforms within the camp are all arranged on the west–east axis. During all other times the sleeping platforms in the camps are all individualized and

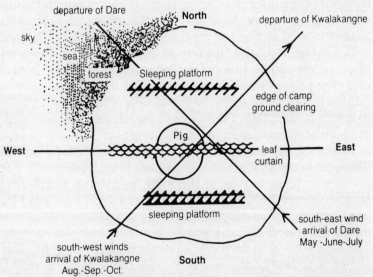

Fig. 11   Direction of Winds, Location of Camp, and Curtain
Under which *Tanageru* is Held

arranged on the north–south axis. The leaf curtain (*enerayale*) is raised in the centre of the camp-ground (enclosed by the sleeping platforms). It forms the line dividing the camp-ground into two, and is along the west–east axis (*see* Fig.11)

The Ongees were very particular in arranging the curtain in this particular way. The explanation for it was:

> We set the curtain the way it is now [on the east–west axis] so that Dare coming [from the south-east] is stopped by the *enerayale* and then gets to find [have *malabuka*] with the *naratakwange*. In the same way when *Kwalakangne?* is about to come from south-west it too has a *talabuka* at the *enerayale* and this way the *naratakwange* comes back if the child fails to come back after the *tanageru*. The leaves of *teigatale* are attractive for the spirits but they do not fall with the winds and spirits so the spirits and winds keep coming to us.

In the above statement made by Totanage, the father of the novice, the purpose of constructing the *enerayale* becomes identical with that of the activities as described in the first *tanageru* myth (Myth no. 6, unit M).

As in the myth, the human beings send the young boy to the spirits and bring him down. The construct of the curtain makes it possible for the Ongees to get the spirits and winds to come and take away the novice. The curtain thereby serves as a 'trap' as well as a 'mask' for the spirits made by the Ongees, which makes it possible for human beings to have the desired effect on the spirits. Just as the Ongees use fire smoke, bones, and clay paints to avoid conjunctions of movement between human beings and spirits, the *enerayale* is made by the Ongees to have a conjunction of the paths of the winds and spirits with the path of the novice sitting down 'to go up'.

Once the initiate is placed under the curtain, the process of going up becomes a possibility and not just a probability. For the novice to go up, as in the myth pertaining to the first *tanageru*, the probability has to be changed into a possibility by transforming the body of the novice in accordance with movements generated, and stopped within the whole world-view. The *naratakwange* is devoid of any substance (like the nautilus shell). His body, deconstructed by the women through *enatatanuye*, is under the curtain, ready to be processed further.

Chogele-gababetebe *and* Galebey:
Feeding the Initiate with Blood and Hanging Fish

Through the process of *enatatanuye* the women have transformed the initiate's body. 'The initiate's body is light and
devoid of smell', is the Ongee interpretation of all the activity
undertaken before the initiate is brought to sit under the leaf
curtain. After the initiate is placed under the leaf curtain 'smell
in the body is put in'. This implantation of smell in the body of
the initiate is called *galebey* and is achieved when the initiators
feed the hunted pig's blood to the initiate. Thus the basic idea
in performing *galebey* is 'giving' the initiate the smell of pig
(Myth no. 6, unit E).

All the camp-mates come to the initiate sitting on the pigs.
The father and the initiator sit on the novice's side, supporting
him from the back and with the other hand start feeding the
pig's blood into the novice's mouth. The leaf curtain is so low
that nobody but the father and the initiator can see the novice
feeding on the blood. Often the novice vomits out the blood.
However, the initiator and father take care that nothing is
wasted in the process of feeding blood to the novice. The blood
that is rejected by the initiate is collected and put back into his
mouth.

Feeding the blood of the pig is a time-consuming process,
since none of it is to be wasted. Every time the novice succeeds
in consuming a handful of pig blood the men standing around
the novice hang *chogele* (leaf ornament) on the leaf curtain.
Chogele originates from the term *choge* which means fish, and
the suffix *le* makes it 'just fish'. The hanging of 'just fish' on the
leaf curtain is called *chogele-gababetebe*. *Chogele* are of two distinct forms, representing two different kinds of fish. Straight
stems about twelve to fifteen inches long with one end cut to a
sharp point are representative of needlefish—and are called
teralu chogele. The other form of *chogele* is made out of the bark
of areca palm, shaped into oval disks with a fish-like tail at one
end. This form represents *tegure* (tuna).[12] The Ongees explain

---

[12] *Chogele* are tied into pairs so that they are easy to hang on the leaf curtain. I did
observe on some occasion when single *chogele* were tied to the leaf curtain by the
men while the initiate was being served blood. *Chogele-gababetebe* is performed only
by men and not women. Women stand watching the father and the initiator feeding

the hanging of the 'just fish' on the leaf curtain in the following manner:

> Doing *tanageru gikonetorroka*—we are all taking all those things that we are not supposed to take away from the spirits. Taking away the honey made all the places dry and devoid of spirits, and now we have to take something else so that again the spirits will be angry— instead of going away they will now come due to anger and *injube* will remain no more dry as it was after the Torale. We all have started hunting the full-grown male pigs from the forest for the *tanageru* and in return what we are letting the spirits take is the *naratakwange* (novice). So we have to stop the spirits at our camp-ground, after the pigs have been hunted by the novice, we make the leaf curtain—spirits cannot see well, they smell and feel things—so to remove his confusion about why he has found the initiate under the leaf curtain we hang the *chogele*—it is all *akwabeybeti* (bad and offensive work).

The above statement was made by the novice's mother's brother, Taiee. Taiee along with the other Ongee men, regarded the hanging of *chogele* as an act within the chain of offensive and bad acts in the course of the ritual. For the spirits, *chogele* is a confusion created by the Ongees, based on the inability of the spirits to see and only are able to smell and feel things by touch. When the Ongees hang 'just fish', they present the spirits with something that is a substitute. The Ongees are aware that the needlefish and tuna fish made from plant substances present the spirits with something that is not fish. For the Ongees, this is a way of telling the spirits that the Ongees have consumed (in terms of hunting and gathering) what they were not supposed to. As the Ongees put it, 'we cannot find the fish [*teralu* and *tegure*] that spirits want us to collect and feed on!' This aspect of Ongee communication to the spirits has its basis

---

the novice. It is significant that needlefish (*teralu*) is the least preferred fish in the Ongee diet. Often the fish is rejected and discarded in the hope of getting *tegure* (tuna). Tuna is the most preferred fish in the Ongee diet. Value attached to the two different fish is so strong that Ongees say, 'I am so hungry now that I would even eat *teralu*!' On many occasions when I would be eating various kinds of fish brought in by the camp and express my appreciation, Ongees would say 'Yes, it is all fine but we did not get any *tegure* today!' Ongees find *teralu* easy to shoot with arrows since they are found close to the shore and are seen all the year round. *Tegure*, on the other hand, is a fish which is 'found away from the shore and difficult to shoot with arrows'.

in the myth of the first *tanageru* (Myth no. 6, unit M). This indicates to the spirits that the Ongees cannot find the food that the spirits had asked them to consume, after entering into a 'give and take' relation with the spirits. The purpose of this agreement between the Ongees and the spirits was to make sure that the Ongees would not hunt pigs, and if they did then the angry spirits would come down and take away Ongee men. Thus, by hanging 'just fish', which is preceded by the initiate killing a pig in the forest, the Ongees achieve a two-fold purpose. They attract the spirits to come and take away the initiate sitting under the leaf curtain. Second, the spirits who come to the island and take away the initiate, in accordance of the myth of the first *tanageru*, also end the seasonal duration made by men (i.e. *gengeyebe*).

In the myth of the first *tanageru*, the events pertaining to the initiate, the spirits' movement, and the conditions imposed on eating fish are as follows:

Event A: The spirits bring down all the males.
Event B: The Ongees are instructed by the spirits to eat *teralu*.
Event C: If the Ongees do not eat *teralu*,and hunt pigs to eat,
Event D: the men would be taken away by the spirits.

Within the mythical account of the first *tanageru*, the events A and B mark a situation where the spirits start a seasonal duration by coming down to the island with men, winds, and rains. However, the ritual of *tanageru* reflects a different ordering of the events. When the Ongees perform the ritual of *tanageru*, they first kill the pigs (event C) that was proscribed by the spirits. The outcome of this is that the spirits become angry and come to the island. To make sure that the angry spirits realize why the pigs were killed, the Ongees erect the leaf curtain and hang 'just fish'. By starting with the myth's event C, the Ongees succeed in creating event D within the ritual sequence of *tanageru*. The spirits who come down take away the novice sitting on the pig under the leaf curtain. According to Teelaie, one of the respected elders among the Ongees:

In doing *chogele-gababetebe* we tell the spirits that we looked but did not find *choge teralu*, neither did we find *choge tegure*, the fish we like most! You see it is not what we would like but the fish is not in the sea. On not finding any fish we show to the spirits *chogele*. It is not to tell a lie or trick the spirits. We do all that like throwing wooden phalluses, getting honey that are tricking, and telling a lie to the spirits. But hanging *chogele* is different. It is the truth! It was *gengeyebe* because we could trick the spirits. Now it is about to become *chongojebe* and to make it we have to tell the truth to the spirits. It is no good to keep tricking them to come to us—to return our children and give us more children.

Babagegi, the chief initiator (*muteejeye*) of Totanage's son took the explanation a step further and said,

> We never find *teralu choge*, we would eat only when nothing is around—we do not find *tegure* the most liked fish—since we cannot see it we cannot have it—they are all away and would be around after the Kwalakangne? leaves our *injube*—we have to show this all to Teneyabogalange and Dare so that they know we want to end everything—we want them to come. So we show them that we killed pig—have my *naratakwange*—but see there were no *choge* so we show only *chogele*!

From Babagegi's point of view, the *chogele* placed on the leaf curtain makes it clear to the spirits that pigs had to be killed since no fish was to be found.

Hanging of *chogele* not only is a reordered presentation of events and categories, but is an apology made by the Ongees to the spirits. By hanging the *chogele* the Ongees acknowledge the relationship between men and spirits, which makes it possible to have *chongojebe* as well as *gengeyebe*. It is *chogele* instead of *choge*. It is a statement that is more than 'symbolic' or 'sentimental' in nature.

The fact that the hanging of *chogele* is not just a symbolic act has some bearing on how the Ongee myth of *tanageru* and the ritual of *tanageru* become related. This relationship between myth and ritual is of reflexivity, and the two do not just mirror or distort each other, and are not merely dialectical and anti-structural with respect to each other. Radcliffe-Brown, in a chapter dealing with Andaman beliefs, myths, and legends, says that the relationship between myth and ritual is that of 'symbolic thought'. He compares the often inconsistent, concrete, and metaphorical symbolic process to the process found in dreams and in various forms of art (Radcliffe-Brown, 1964: 397). For Radcliffe-Brown the purpose of the myth is to express sentiments and desires of various kinds, 'the *either–or* relation is inadmissible owing to the very nature of the thought-process itself' (Radcliffe-Brown, 1964: 396). Although in day-to-day existence and practical aspects, the Andamanese are 'excellent observers of natural phenomena and are capable of putting their observations to practical use', their myths do not follow the laws of logical reasoning. The explanation, Radcliffe-Brown suggests, lies in a kind of psychological law governing the use of symbols to express emotions:

> . . . a mind intent on expressing certain feelings, faced with two alternative and equally satisfactory but inconsistent symbols, will hesitate to choose between them even at the command of the desire for logical consistency. It will cling as long as possible to both of them. [Radcliffe-Brown,1964: 396.]

Andaman mythology follows such a psychological principle:

> The view of lightning as a person who shakes his leg seems to express in some way certain notions of the natives about lightning. The alternative explanation of lightning as a fire-brand thrown by Biliku also satisfies in some way his [the native's] need of expressing the impressions that the phenomena make upon him. In spite of the inconsistency he clings to both symbols as best he can. [Radcliffe-Brown, 1964: 396–397].

According to Radcliffe-Brown's observations and formulation, the 'symbols' in this case are personifications of a natural phenomenon that appeal 'through the imagination, to the mind's affective dispositions' (ibid.). By using such symbols and metaphors, Andamanese myths express the social values of proper conduct, of the past and tradition, of familiar places and landmarks, of weather and seasonal changes, and things found in the forest. Taken together, the myths express a system of social values or sentiments upon which Andamanese society depends for its existence. The myths represent for the Andamanese a moral and material order of the universe, that is, a cosmology, fragmentary and inconsistent but capable of systematization by the anthropologist. Radcliffe-Brown's conclusion from his symbolic interpretation of Andamanese myths is that their function is Exactly parallel to that of the ritual and ceremonial' (1964: 405).

Radcliffe-Brown's idea of the 'symbolic' and the parallel

between myth and ritual is very different from what the Ongee individuals do at the time of *tanageru* and how they relate to the myth of the first *tanageru*, especially what is reported by Taie, Teelaie, and Babagegi in relation to the hanging of *chogele* on the leaf curtain. The ritual and myth may have parallels but the relationship is reflexive. This is so because the relation of 'give and take' (*gae?bebe*) between the Ongee and his spirits is responsive in terms of the formation of *chongojebe* and *gengeyebe*. The responsive nature of the relation between man and spirit is reflective of notions of offence, precaution, *malabuka*, and *talabuka* in a shared space. The Ongees are aware of this reflexivity between the myth and the ritual because they are 'excellent observers of natural phenomena' (Radcliffe-Brown,1964: 396) and not 'inconsistent' with the either–or relation' (ibid.: 396). For the Ongee, the performance of the ritual of the wooden phallus, the killing of pigs, and the hanging of *chogele* so that the novice can 'go up to the spirits', are based on a very logical relationship between myth and ritual. The reflexivity of Ongee myth and ritual is based on the principle of the self-reciprocating asymmetrical relationship between the Ongee and his spirit. It is this asymmetry of the self-reciprocating relationship between man and spirit that makes it possible for the myth and ritual to be reflexive, which in turn lends itself to the formations of *injube* and movements in it. This is the context of meta-relations by which the relations and elements of smell, spirits, animals, winds, and human beings are stratified for the purpose of power being transacted, exchanged, given, and taken. It is in this context of meta-relations within the *injube* and movements that the Ongees regard themselves as being constantly hunting and hunted, being born and dying —coexisting with fellow beings and spirits in safety and in danger.

 Chogele-gababetebe marks not only a point in the chronology of events forming the sequence of the ritual of *tanageru* but something more. In the logic of *chogele-gababetebe* lies the change that is brought about in the way the Ongee and spirit relate to each other. After the *chogele-gababetebe* are hung on the leaf curtain the moment of the initiate's departure to the domain of the spirits is a step closer. While the *chogele* are being hung on the leaf curtain, all the blood of the hunted pig is consumed by the novice. Then the chief initiator, *mutarandee*, moves to a position behind the novice sitting on the pig. The *mutarandee* places his hands on the shoulders of the novice and starts the phase of *gemey?be*, form of massage.

Gemey?be: *Massaging the Body to Alter Weight*

Standing behind the novice, the *mutarandee*, with one foot pressing on the back of the neck, starts pressing down on the vertebral column of the novice. This downward movement of the chief initiator on the back of the novice is an essential way of 'pressing and pushing' (*gemey?be*) the blood consumed by the novice. The 'pushing and pressing' (*gemey?be*) forms the core of the Ongee idea of massage for the care of the body. In various contexts, with various substances, the human body is massaged for the cure of its pathological condition (Cipriani, 1961).

In the course of the *tanageru* ritual, the blood consumed by the initiate should settle down and inside the novice's body. Achieved through *gemey?be*, this is important because it is extremely likely that the novice's body may reject the blood served to him. The blood should remain in the body of the novice, since the novice consumes nothing else, which makes his body light in weight and devoid of any smell. The blood served to the novice and the body massage make it possible for his body to acquire some weight. This addition of slight weight enables the spirits to take the lightweight novice up to the sky and then return him. The Ongees believe that the body has to be light enough for the spirits to take the novice away, but his body should also have some weight to enable it to return from the spirits' home. For the Ongee, this idea is embodied in the nautilus shell which floats but never sinks like other empty shells. Thus, the initiator's explanation for the massage is, 'What we are doing is making the novice into a nautilus shell with the smell of pig—this would make it possible for the boy to go up to the spirits and talk with them and then come back to us'.

After the chief initiator has massaged the novice, all the other camp-mates do the *gemey?be* on the novice's body. In doing *gemey?be*, an order is followed by the Ongees. First, all the women related to the novice matrilineally do it, then come all the men related to him maternally. Then all the women and men related to him patrilineally massage him. In the end, young children are helped by their parents to massage the novice's vertebral column. In this way everybody contributes in pushing down weight in the novice's body. Everybody here includes all those who will eat the pig hunted by the novice as well as those who will not.

Once everybody has massaged the initiate's body, the *mutarandee* cuts the *atakee* (garment made of bark around the waist of novice). The *atakee* is then hung on the leaf curtain. The arrow with which the novice has killed the pig is held in the left hand of the initiator, and the novice is vigorously shaken by the *mutarandee*. The shaking of the completely naked novice sitting on the pigs is a sight that is not to be seen by the women. As the women leave, the shaking of the initiate becomes more and more vigorous. This is known as *gambotukweg?-chuwebe-eneyobochuyebegi*, and literally translates into, 'wake up, get hurt like the arrow hurts the hunted body and do not forget to listen carefully'. The act of vigorous shaking by the initiator culminates in a quick recitation known as *gekonene* (literally, do not look for it). It is the significance of the *gekonene*, a short and loud recitation by the initiator, that requires the vigorous shaking.

When the initiator recites the *gekonene*, the initiate is told of either a food item or an object made by human beings that would become *gilemame*, proscribed to the initiate, after the initiation is over. The proscription (*gilemame*) refers to the individual not consuming the specific food or use of that particular object for himself. After the *gilemame* is declared, the individual is required to only procure the specified food for others and make that particular object only for others. The declaration of *gilemame* in the discourse of *gekonene* by the *mutarandee* is called *yen?dangebe*.

The following items were enumerated by the Ongees as things that can become *gilemame* for any and every individual male within the Ongee society. Each of these items, on becoming *gilemame* and in the context of *gekonene*, is referred to by a ritual term. Also, the initiate throughout his lifetime refers to the *gilemame* by the ritual term.

OBJECTS OF AND FOR *GILEMAME*

(1) Ongee term	(II) Ritual term	(III) Meaning	(IV) Associated Season
Nana	Tenyaboro	Prawn	August Kwalakangne?
Tombowage	Tobokale	Cicada	October
Tambanua	Naretorangele	Pig	
Tambanua	Kanameyakange	Pig	of Mayakangne?
Tambanua	Kanadare	Pig	of Dare
Naralanka	Torantono	Turtle	November–February
Choge	Nagecele	Fish	February–March
Tuouwe	Tenyabone	Dugong	July–August
Dange	Nareterabu	Canoe	April–May
Ukku	Narelauehye	Honey container	February–March
Toneyage	Oyetabawa	Adze	January–February
Lehye	Koyekwale	Knife	October
Aa	Narageru	Bow	July
Chenekwa	Centanarere	Arrow	May
Cekwe	Gejekabe	Fishing net	September
Tole?ye	Naretandaleye	Basket	June
Ebolekubwe	Tomegeru	Leaves woven together for thatching	August
Boneye	E?oreuwe	Resin	December–January

There is a particular logical way for the Ongee to select one of the *gilemame* from the above list. The *muteejeye*, the chief initiator, who is also the *mutarandee* at the time of initiate's birth, selects the object of proscription depending on the season in which the child was born. Thus, in recalling the season of birth for the male child (the initiate), the *mutarandee* (now the initiator) acknowledges the initiate's relation to the spirits and to his fellow human beings. In the selection of *gilemame* we can also see the Ongee consciousness of the link between child and spirit.

As the male child grows, feeding on the milk that all the women of the camp provide, the *mutarandee* is informed when all the breast-feeding mothers realize that the lower and upper teeth of the child have appeared. The child is regarded to be 'now a human' in the season in which the *mutarandee* is informed of this development. He takes note of the particular season in relation to the child's development by the food item available to the camp and things that are made in the camp. Generally the *mutarandee* keeps quiet about this, but as the child grows he makes sure that the child has learned as much as possible about the specific object manufactured or food item procured during the season which he has indexed as the point when the spirit has become a human. No one talks or tells the novice about this while he is being brought up within the community of the Ongees. It is only at the time of the *tanageru* that the *mutarandee*, who has become the *muteejeye*, tells the novice about what is to be a proscribed item (*gilemame*) for him.

Once the initiate has been told of his *gilemame* he is expected never to refer to it by the term used for it in day-to-day usage. The initiated man is only to refer to the object by the ritual term that is identified in the course of the *gekonene* (*see* col.2, p. 241).

In the change of term of reference for the *gilemame* what is important for the Ongees is that it makes it possible for all to understand that the specific food item or product is something that the individual may not use but will nevertheless provide to others. The existence of two terms of reference for things that may become *gilemame* makes it possible for the Ongees to make practical use of a highly valued characteristic trait in each individual man's personality. Totanage, father of the initiate, who never eats prawns since it is his *gilemame*, said,

> One should not use his *gilemame* for himself—it is like marrying brother and sister, *gilemame* should be procured and made but given to all others around you just like giving to the spirits, to make possible returning back after the *tanageru*. It is important that *gilemame* should not be forgotten about and one should give *gilemame* to the others. After my *tanageru* prawns (*nana*) became *tenyaboro* for me.

In this statement of Totanage there is a correlation made between marrying sisters and brothers with not exchanging (give and take) between men and spirits. In Ongee language, improper marriage alliance, incest, one who is selfish, and one who does not share with others are all referred to by one term, *ma-gae?bebe-ma*, literally meaning one gives nothing and takes all or one who keeps all.

It is the range of meanings associated with *ma-gae?bebe-ma* that Ongees perceive a breakdown of give and take relations between man and spirit and between man and man that could be referred to as the Ongees' highly valued virtue in a person as self-denial and the community's indulgence. Thus the novice hunts and feeds all but remains as an individual proscribed from eating meat throughout the *tanageru* ceremony. In the same way after the *tanageru* the individual initiated is expected to make and/or procure the *gilemame*, be it a food item or a *maonole* for others but not for himself. It is this principle of refraining for the self but providing for the larger collectivity that fosters goodwill and interdependence within the small band organization and a small tribe limited in resources on an island where there is competition with the spirits and an irregular supply of subsistence. Denial for the self leads to positive results and is therefore a demonstrable truth for the Ongees and the individual because an individual by telling *gilemame* to the spirits cannot be detained by them when he visits them during *tanageru*. In the return of the individual, after he has promised to provide *gilemame* to the spirits, the community gets back the individual as well as the individual's role of providing the same *gilemame* to his human relatives. It is in this *gilemame* that the relations as well as relations of exchange between humans and humans and spirits get objectified.

In the course of the initiation ritual the initiate is to visit the spirits and return. But at the spirits' residence the initiate is expected to inform the spirits about the *gilemame*. The Ongees believe that telling the spirits about the *gilemame* is an individual's promise to provide it to the spirits.

According to Totanage, father of the initiate:

My son has to know his *gilemame* well and remember to tell to the spirits as well remember it when he comes back from his visit to the spirits during his *tanageru*. The spirits will let him go only if he knows *gilemame* because my son will give the *gilemame* to the spirits whenever they want it—see, the spirits can then count on help from the specific individual in terms of getting *gilemame* from him since they at the time of that individual's *tanageru* let him return. My son when he tells his *gilemame* to the *tomya* during his *tanageru* also makes the spirit know that one of them has become an Ongee and it is not a false thing because he has learned everything that Ongee men do and knows also one thing very well. See, the spirits do not like that—they cannot hunt, they enter whatever they want to feed upon—but my son will help them by giving food specific if it is his *gilemame*. Spirits cannot even make *maonole* (tools) so if the *gilemame* is a specific *maonole* the spirits know that my son is no more a spirit because he can kill and make *maonole* and they [the spirits] can get it from my son!

Thus the communication of the *gilemame* by the initiate to the spirits makes it possible to distinguish between one who was once a spirit but is now a human (i.e. the Ongee initiate) and one who was once an Ongee but is now a spirit. The negotiations and agreement between the spirits and the initiate, with a focus on the *gilemame*, make it possible for men and spirits to coexist and sustain their give-and-take relations. The 'give and take relationship', *gae?bebe*, has its importance within the Ongee culture, and indeed has a significance greater than that of a 'total social fact'. The Ongees explain their existence and origin since mythic and historic time through the give and take relation (see Myths no.1, 2, and 4). Even the condition of decline in the Ongee population was expressed by Raja Nappikute?ge using the idea of *gilemame* within the give and take relationship as follows:

After our ancestors started getting too many things from outsiders and started going to Aberdeen [Port Blair] we had many wars— many died—we started forgetting about our *gilemame* — more and more spirits got angry because we were not giving them anything. Outsiders started to live in our places—no food so the angry spirits started taking away Ongees. Now we are so few left. If we do not give to the spirit the *naratakwange*, if the *naratakwange* does not give *gilemame* to the *tomya* as well as Ongees, then all of us will keep dying, there will be no Ongees. Just *tomya* and *tomya* — nobody to marry, no give and take, nobody will be here only *tomya* and outsiders. Then *tomya* too will go away forever because outsiders do not give to *tomya* as we do to them. The *tomya* too will die and only outsiders will live. We have to give and take, especially a man has to know his *gilemame* otherwise all the games of 'hide and seek' will come to an end.

Gilemame, which articulates the 'give and take' relationship for Ongee society, is of great importance. The initiator in his capacity as a *mutarandee*, imparts all his knowledge and skill to each Ongee boy, but in this training what has to be declared as *gilemame* in the course of the *tanageru* is given special attention. While the *mutarandee* shows and says all about what would become *gilemame* for the boy (in the course of *mutarandee* reciting *gilemame* to the initiate), and in his capacity as *muteejeye* (to the boy) he makes it clear that the categorical reference to *gilemame* is a form of knowledge to be practiced. A knowledge that the novice has to remember, for on this depends his existence and capacity to move.

This aspect is put forth to the novice sitting on the pig by the *muteejeye* in the following manner:

> Go go now now you have to go
> Fear not Remember you have to come
> Give what is required
> But give when the spirits come
> Spirits come, winds come, you come
> Listen well!
> You are xxxxx
> You are not yyyyy
> Quick will end the coming and going
> If you remember and tell the spirits.
> What to give—
> To give is to take
> Go go now now you have to go on
> Fear not
> We will wait for you, remember remember xxxxx.[13]

As the *muteejeye* finishes the lines of recitation, the novice's mother rushes towards the initiator and snatches the arrow from his left hand. The mother of the novice detaches the

---

[13] The recitation of *yen?dangebe* with the *gilemame* framed between showed little variations in the two cases of *tanageru* which I observed during the course of field-work. Within the recitation done by the *mutejeeye* the term from column 2 is inserted in place of 'xxxxx'. 'xxxxx' stands for the place where the ritual term for the *gilemame* is inserted in the course of recitation. In the same way, in place of 'yyyyy' the *muteejeye* inserts the equivalent of x from column 1, p. 241.

arrowhead from the shaft and throws it over the leaf curtain.
This marks the end of a stage, but also marks the beginning of
the one where the novice is about to start his journey to the
spirits' residence.

The men lift the novice from his seat of pigs. The novice's
body, practically lifeless in appearance, is brought to the
women who have all gathered at the extended sleeping platform. For the next two days, known as *eneyetandenetakwebe*, all
the women will take care of the novice's body. The Ongees
believe that it is in the duration of these two days that the body
internal goes up to the spirits' home, and only the body external
is left with the camp-mates. During *eneyetandenetakwebe* the
novice is referred to as a 'nautilus shell that has floated away'.
Before the novice's body internal departs, the women start
'heating' up the body external and continue heating it. Only
after two days is the novice served with cold water and is then
regarded to be one who has come back.

Heating the Body to Meet with Teneyabogalange

As the novice's naked body is brought to the women, the
processes of heating and massaging start. This is a prerequisite
for and prelude to the novice's trip to the domain of Dare and
the spirit of Teneyabogalange. Heating and massaging are
particular means of making the 'body internal' move out of the
'body external'. The novice is laid on the laps of the women
who sit facing each other in two rows. The only woman who is
not allowed to sit here is the novice's *angachee* (wife) because
she has a special role to play. She stands close by, waiting for
instructions from her mother and her mother-in-law. The *angachee* is to provide small smouldering pieces of resin wrapped
in green leaves to the women on whose laps her husband's
body is lying. The wife's own mother is the first to receive the
resin-wrapped leaves. The wrapped leaf with smouldering
resin inside it, which forms a heat pack, is pressed against the
novice's body for a short while and passed on to the woman
sitting next in the row. In this way all the women around the
novice get a chance, one by one, to apply the heat packs. Each
woman in this way applies heat to the part of the body of the
novice that is in front of her. As the heat packs get passed

around they get cooler, and finally the novice's mother (*anga-chee*'s *meeragele* [mother-in-law]) hands over the used up heat pack to her *mambetageye* (daughter-in-law). Throughout, the *angachee* keeps busy preparing new heat packs.

The heat packs (*boneye?alogeye*) and how they are placed are important and so is the wife here, who alone is entitled to prepare them. As the Ongees say, it is the 'wife alone who knows how much "heat" is required for cooking food that the husband can consume'. Also it is the wife alone who applies the clay paint on her *magechebe*'s (husband) body, which is also related to aspects of heating and cooling. Thus, before a man's *tanageru*, it is important that 'he should be married so that his wife could contribute to the important process of wrapping and packing of *boneye?alogeye*. The importance of the massage and placement of heat packs lies in their being seen as a form of 'cooking' the blood inside the novice's body, and lead to the hardening of smell within him. The placing of heat packs and massaging the body is called *achengekwayebe*, which literally means pressing in heat on the smell.[14] According to the Ongee women (who alone are entitled to perform this kind of massage), the heat, upon being pressed in on the smell, hardens and cooks the smell within the body. The raw blood that has been given to the novice especially needs to be cooked and heated, otherwise the novice's body will have no smell and will feel cold. Also, failure to heat up the body after the consumption of blood may cause the novice to vomit it out. The other reason for performing *achengekwayabe* is that it is a way in which heat is transplanted by the women into the body of the novice. This is essential, since the novice is going to experience intense pain and the sensation of shivering from cold (*lololobe*) when he meets with the spirits. After a few rounds of placing the heat packs and massage, the novice's body is smeared with red clay mixed with honey and melted beeswax. This paint is called *e?uetalukene*, which is the verb for retaining something. The honey and beeswax mixed with the red clay paint, which by its characteristic is a heat-generating substance, are a way of keeping the red paint from drying up and peeling off from the body.

[14] This form of massage and heat application was observed even when Ongee women were going through labour pains, childbirth, and on the body of the new mother as well as new-born child immediately after birth.

However, when the red paint is applied with the honey and wax, it is regarded by the Ongees as forming a sheath which checks and slows down the loss of body heat. In other words, the honey and wax are seen as covering the red clay paint, which prevents the heat of the paint from being lost. This makes it possible for the novice's body to remain warm.

As the red paint with the honey and wax mixture starts to dry, the women start tying thread, made from the fibre of hibiscus, around the face and around the limbs and waist of the novice. All the households contribute to this thread tied around the novice's body. The tying of thread is known as *eneyee?e-boketkabe* and is seen as a measure that prevents the body external from falling apart when the body internal returns.[15]

A cane stem, about four inches long, is hung around the novice's neck. This 'tube' with a decorative design woven on it with strings, is called *umma*. The *umma* serves as a suction tube by which the novice consumes all the liquid food served to him towards the end of the ritual. The reason given by the women for the *umma* was:

> Our child is dead, he sucks like the spirits, when he comes back as a child after two days and two nights he will ask for something, he has to be served something that he can consume through the *umma* and till his body has grown up to deal with weight; around the time it is proper to end the ritual of *tanageru* he will give up *umma* and start eating food since his teeth will be back at work.

Now the novice is ready to leave for the visit to the spirits. His body is covered with a sheet of cloth known as *guyerotukey*, although traditionally the Ongees used to cover it with just leaves.[16] Under the sleeping platform, a small fire is also set up to keep novice's body external warm till the body internal returns. For two days and two nights the novice remained covered and in the custody of the women who took turns in sitting around him. Occasionally the women would turn the

[15] This may be correlated to the idea of the Ongee women tying strings to the body of the child in the myth of first *tanageru* (see unit K of Myth no. 6). However, none of the Ongees see any connection between the tying of strings around novice's body and this, and say it is related to the myth of the first *tanageru*.

[16] During my field-work my own ground mat was used, since it was regarded as being very strong and capable of generating heat due to its thickness. Ongees also used the plastic sheets I had along with me to cover the novice.

novice's body from side to side. Every time the novice sighed or tried to talk in the stage of 'body external asleep' and 'internal on the move with the spirit', the women would comfort him and say, 'He is there where the spirits are! he is, he is!'

This duration of two days and two nights is to end on the third day's evening when the novice, on making the first sound of waking up, is served a mixture of warm water and leaves of *Heritiera littoralis* and *Hibiscustilaceus*, which the Ongees regularly drink and call *muroie*. *Muroie* is served to the novice in a nautilus shell and the women help him to drink it through the *umma*. The duration when the novice is taken care of by the women is called *eneyetandenetakwebe*. As the initiate's body is processed by the women, all the men raise the leaf curtain higher and start cooking the pigs hunted by the novice, which involves singeing, cutting up the animal's body, and finally setting all the parts to boil. The pig is cooked in a cooking area that is specially designated for the *tanageru* ceremony. This kitchen always has to be outside the main camping ground (*wabe*), and is set up behind the lines of *korale*. Unlike other situations, the raw meat of the pig killed during the ritual is never distributed around the camp. Only the cooked meat is distributed among the camp-mates. All those who eat the meat are very careful not to throw out the bones and the leftover meat. What is left (bones and scraps) is brought by all the individual consumers to the kitchen. At the kitchen, which is managed by the wife of the initiator, the bones and scraps are collected in a basket and hung on a nearby tree. This rule is to be maintained till the ritual of *tanageru* is over. In all day-to-day situations (i.e. in a non-ritual context), the bones and scraps of pig meat are thrown around the camp-ground and the packs of dogs clean them up.

For two days and two nights all the Ongee camp-mates spend time eating, resting, and taking care of the novice's body external. On the third day, with the sunrise, slowly and gently all the covering of leaf and cloth on the novice's body are cut loose. With the help and support of all the women the novice is made to sit up. He is served *muroie* in a nautilus shell, which he consumes through the *umma*. Throughout, the novice keeps his eyes closed.

This marks the end of *eneyetandenetakwebe*, the novice's visit to the spirits' home. On his successful return, *edatamakwbeye?e*, from the home of spirit Dare, the Ongees start on the second phase of the *tanageru* ritual. This mainly serves to deal with the aspects of further transforming the body that has come back from the spirits' domain, and to celebrate the return of the novice.

CHAPTER 7

Phase Two of the *Tanageru* Ritual

After two days the novice has returned from his visit to the home of Dare. For the Ongees, this marks the success in sending up the novice and his return is now to be dealt with. For the Ongees the person who has come back (i.e. the novice) is not the same as the person who was processed and sent up to the home of the spirits. The novice who has returned is regarded to be a body sharing its identity with a spirit and is much like a child whose body is to be worked upon. This working upon the novice's body pertains to cooling it and distributing the weight in it. For the whole community, the return of the initiate (*naratakwange*) is a reason for celebration, since it implies that the spirits have agreed to come back, and that the seasons caused by the movement of the winds and the spirits will start again. Once again it will be a domain where the spirits will act and humans will suffer the impact. Once again children will be born.

On the morning of the third day, after the novice has consumed *muroie* through his *umma*, the women start to massage his body again. His body, completely naked, is very vigorously massaged. This massage is called *gamyebe*. The *gamyebe* is distinguished by the Ongees from other forms of massage performed after feeding the novice the pig's blood. The others are in principle regarded by the Ongees to be 'pushing the weight downward till the waist and then the lower half of the body is dabbed upon, as if the weight is to be kept there'. This, according to the Ongees, is a form of *kayareye*, which literally means low tide. The Ongees associate movements of smell and breathing with the tidal movement, and see the massage too as a way of shifting the body smell and the weight downward, forming *kayarye*, low tides.

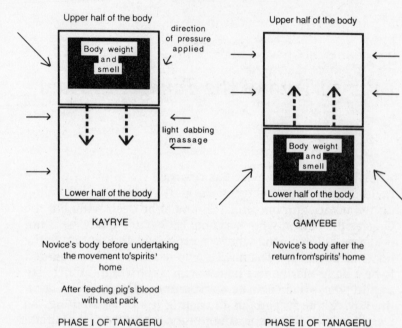

KAYRYE

GAMYEBE

Novice's body before undertaking
the movement to 'spirits'
home

Novice's body after the
return from 'spirits' home

After feeding pig's blood
with heat pack

PHASE I OF TANAGERU   PHASE II OF TANAGERU

**Fig. 12 Two Distinct Forms of Body Massage to Alter and
Transform the Novice**

After the two days and two nights that the novice has spent
visiting the spirits, he is regarded to have returned as light as
a 'nautilus shell', which enabled him to float and undertake the
movement to the spirits' home. Once the nautilus shell has
come back, the lightness of the novice's body is to be changed.
It is for this reason that the massage now done is *gamyebe*, which
literally means high tide. The pattern of this massage is the
inverse of the *kayareye* form of massage. *Gamyebe* is undertaken
by an upward movement of the hand pressure from feet to the
waist and by squeezing on the upper half of the body. Totanage
and Tambolaie, whose sons were undergoing *tanageru*, explained the *gamyebe* massage in the following way:

> Women are now trying to push the smell and weight from the lower
> half to the upper half of the body—so that the returned nautilus
> shell can remain safely and heavily with us!

Schematically the two forms of massage are represented in
Figure 12. In this analytical representation, I have tried to show
the two distinct stages of the novice's body transformation
accomplished by the society.

After the course of *gamyebe* massage, the novice is adorned
with a new strip of bark around the waist, and lifted by the chief
initiator and brought to the centre of the camp-ground. Right
under the leaf curtain, a few men support and make the novice
stand up. The stable standing of the weak novice is assured by
his father-in-law, who stands right behind the novice. With this
formation starts the steps for cooling the initiate. These steps
are called *yulelale-lale*, and in conducting them the chief initiator plays an important role once again.

Yulelale-lale: *Cooling the Novice's Body*

Yulelale-lale is the name given by the Ongees to a segment of
'enactment' involving certain choreographed movements. It
involves not only the novice, whose body is being processed,
but also the chief initiator, the father, the father-in-law, and all
the other men of the camp as processors of the novice's body.
'Steps of cooling the novice' are regarded as the exact opposite
of what the Ongees had done in the earlier phase of the *tanageru*
ritual. In the first phase the novice's body was heated to make

it light, so that it could 'go up'. Now, through the steps of
yulelale-lale, the body of the novice is cooled so that it remains
down among the Ongees. In the cooling of the novice's body,
the weight of the novice is increased, and this makes his body
remain below with the Ongees, since the increased weight
makes it difficult for the spirits to lift him and take him to their
home 'across the sea—above the forest—beyond the place
where only winds alone can move!'

Water is sprinkled on the novice's face and his eyelids are
forced open by the chief initiator. From this point onwards the
restriction on the novice of keeping his eyes shut is lifted
(however, the novice keeps his head bent down). The considerably swollen feet of the novice are also examined by the chief
initiator. The blisters on the novice's feet are burst open by
using a thorn or a fish-bone. As the blisters are pierced the feet
are sprinkled with water. Care is taken to ensure that all the
fluids that come out do not fall to the ground. As the blood and
pus start trickling down, they are cleaned by the initiator and
smeared all over the feet. The examination of the novice's body
is called *gerabale*, and it comes to a close when the novice is
served a nautilus shell full of green coconut water. As the
novice starts consuming his drink with the *umma*, the novice's
mother shouts out a question: 'Is *naratakwange* [novice] heavy
enough and cold now—will he move with us? or go away!—
What was that made him go away from you all—Do you
know?'

As the mother's question ends, a chorus of women repeats
it loud and clear. In response to the women's question, the
novice's father sets out for the cooking area where the remains
of the pig hunted by the novice had been boiled for the last two
days. The skull of the male pig, with its curved tusks and
devoid of any skin or tissue, is pulled out by him. He then
places the skulls in front of the women and says, 'Here is the
goterange (skulls) for which we have now done with *gerabale*
[examination of the novice's body]'.

During the ritual of *tanageru* all the skulls of the pigs hunted
by the initiate are given to the women who weave cane around
them in the form of fine baskets, and hang them on a wooden
pole known as *naratakwange-ye gotterange tulukucheye*. Meanwhile the novice is adorned with *owetorakabe*, bands of pan-

danus leaves. The light greenish-yellow strips of pandanus are covered with alternate stripes of red and white clay paint. The father, who makes these ornaments, places the strips of leaves on the novice and attaches their ends with twigs and or fishbones. Once placed on the novice's body, the *owetorakabe* are referred to as *narakware*. *Narakware* is also the Ongee name for the red striped design on the chambered nautilus shell. However, the *narakware* made of pandanus leaves by the father, are placed only on the upper half of the novice's body, and that includes marking on the waist, across the chest, and on the wrists and the forehead. The Ongees in their processing of the novice's body clearly show a concern with the upper half of the body, as opposed to the focus of the massage and string ornamentation on the downward orientation towards the lower half of the body.

The Ongees attach great value to the activities of binding and tying, which have a transformative and generative capacity. The Ongees themselves distinguish between the *narakware* tied during the first phase of *tanageru*, which are tied tightly using 'short lengths' of string, and those tied in the second phase of *tanageru*, which are regarded to be loose and long in length.

Short things tied tightly are regarded to be a means of 'slowly cutting' the body and thereby releasing smell which makes the body light. Opposed to this is the idea that longer things tied loosely on the body cause less smell to be released (since the thing tied does not have much impact and only the required amount of smell gets released), by which the body's weight gets equally distributed all over the body. That the bound leaves are also covered with red and white clay paint is seen as a further means of cooling the body. Little smell is thus released and the red paint ensures that a certain amount of smell is released, so that the process of altering body weight becomes possible.

The tying is seen as a means of cutting the smell, which in the context of the ritual of *tanageru* and of ornamentation is also distinguished as *narakware*, short and tight, causing the rapid release of smell and loss of weight. Consequently during *tanageru*, tying *narakware* and releasing the smell attracts the spirits to the novice's body. The lightness of the body makes it convenient for it to go up with the spirits and winds. After the body returns, the loosely tied leaves ensure that the body releases less smell, becomes relatively heavy, and less susceptible to being taken away by the spirits.

As the adorning of the body concludes, the men form a single line on two sides of the novice. The father and the initiator are on either side. All the other men hold each other's hands and form a line. The novice's father-in-law stands behind him, holding him. This line of men is right under the leaf curtain facing the direction in which Dare departs. Once the line is formed, the chief initiator's wife approaches the novice and hangs *geyenene* from the initiate's mouth. *Geyenene*, which literally means heaviness (an intrinsic quality), is actually a circular form created by binding together two sections of pig fat saved from the pig hunted by the initiate.

Enactment of Hunting and the Dance for Cooling

As the pig fat is hung from the novice's mouth (*geyenene*) the men start the *yulelale-lale*, which could be described as a form of highly stylized and coordinated 'spot running'. First, the right foot is put forward and everybody turns towards the novice in the middle of the line. Everybody takes a hop jump and when they land the left foot is brought forward. This step is repeated a few times, till everybody's foot movement becomes coordinated. As this stage is achieved by the group of men, the father-in-law, standing behind the novice, taking note of the stage of coordination, in a loud voice and in an abrupt manner declares: '*Malabuka-Talbuka-Malabuka-Talabuka!*'

As soon as the line of dancing men hear this, they all sit down and respond back by shouting '*Ule Ule Ule! Yulelale-lale!*' In this process of sitting down the novice seems like an individual in the line who is very sluggish, slumberous, and uncoordinated in his movements. He is practically pulled down and then lifted up with the assistance of his father-in-law. As the novice sits down, the line of other men gets up and continues jumping, and waits for the father-in-law to make the declaration again. The declaration and the response of the men between the jumping and sitting down continue for some time.

The procedure is brought to an end when the novice, along with the men, succeeds in coordinating the act of jumping and sitting down. This is the high point of coordination. The father-in-law now stops supporting the novice's body from the back. The dance continues with everybody exhibiting perfect coordination. At this stage of the dance, the father-in-law instructs a couple of young boys to go and snatch the *geyenene* from the novice's mouth. The young boys follow the instructions and do exactly as they are told. Having snatched the pig fat (*geyenene*) from the novice, they run away and start eating it. This enactment or *enatandemetakwabe*, which literally means quick release of smell and occurs at the time of killing or of loss of life. This ends the dance of *yulelale-lale*. The line of men breaks up and the boys distribute and eat the treat they have snatched from the novice.

This 'jumping' up and down (Radcliffe-Brown,1964: 246-7) is not a dance where the novice is coerced to coordinate his movements. The force exercised on the novice by the men is not just to make him associate with the society (*see* Radcliffe-Brown, 1964: 251-2). At the end of *enatandemetakwabe* I was told by the initiate's father-in-law

> What we did is not *onolabe* (dance). *Onolabe* is not done when we Ongees are making seasons—having impact on the spirits. Dance is done when our movements have been successful while the spirits influence and condition all our moves. This is the *tanageru igagame* a *gengeyebe*, when we move and spirits get to move accordingly!

According to Temai, the father-in-law of the initiate going through *tanageru*, *yulelale-lale* is a form of *okiyatoranka*, hunting done for the women 'to show them and the novice about our killing pigs—and making possible the going of the novice to Dare'. From Temai's point of view, *yulalale-lale* is a response to the women's questions raised before the dance and in this response the act of hunting (*okiyatoranka*) gets enacted. Temai went on to say,

> We show the women that we all and the novice went to the forest— we moved in the forest—taking each foot in accordance to *malabuka* and *talabuka* —we moved in every place going up and down—and across—then we killed pig and then we went on to kill again and again.

It was pointed out to me that snatching the pig fat from the novice's mouth is actually 'taking away the pig in the forest above which the spirits are! It also is the end of the novice'. Consequently, the act of the boys snatching away the *geyenene* from the mouth of the novice is called *enatandemetakwabe*, killing/death by the quick release of smell. Under *enatandemetakwabe* the Ongees include not only hunting of the animals by the Ongees but also the 'hunting' of an Ongee by the spirits. This makes the enactment of the killing of a pig also the marker of the 'killing' of the individual who starts *tanageru* with the hunting of a pig.

Koyra (initiate's father's brother) puts the 'killing of the novice' in the following words:

> Death, killing, hunting, and meeting with the spirits—all happen very quickly—you just do not know when it will happen—It is like the wind that can come and just blow so strong that things are all taken by the winds or they change the shape. It happens since winds take heat, smell, and weight. When the things like Ongees are too light then the spirits with the winds are successful in *enatandemetakwabe* and *enegeteebe*. So it is important to kill the *naratakwange* (novice) that has come back from home of Dare. If we do not and increase his weight and cool him then he might be *enatandemetakwabe* by the spirits, particularly Dare who sends the *naratakwange* back because he is right behind the child we had sent up.

The Ongees' enactment of hunting, as distinct from a dance, tells the women and the novice about what has happened, and also ensures certain preventive measures against the spirits taking away the novice. In the enactment, the past (of what was hunted) and the future (of what could be hunted) are brought together. It is this fusion and amalgamation that defines the enactment as distinct from the performative aspect of dance. Thus the Ongees say that *onolabe*, dance, is a movement undertaken by them to state how and what happened on a particular day, and not about what had happened and what can happen again. The novice is told about the hunting through an enactment, in which discourse is created about the future as well as the past; this is made clear in the second stage of the second phase of *tanageru*.

*Stage Two of Second Phase: Telling the Novice about
'Hunting with the Smell'*

The novice (as a child) going to the home of the spirits and the
torale going up to the spirits are identical situations. The identity lies not only in the movement and contact set up with the
spirits, but also in the possibility of death as a potential outcome of the meeting with the spirits. This is so because the
novice and the spirits' communicator have to take the risk of
not being able to return. However, there is also a major difference between the *torale* and the novice.

When the *torale* returns, he has to remember what the
spirits told him so that he can tell it to the other Ongees. In the
case of the novice, before he starts his trip to the spirits, he is
told by everybody to remember *gilemame*. The novice, on reaching the spirits, tells them all that he has been told to remember.
When the spirits are told this by the novice, it becomes possible
for the novice to return; also the chances of the winds and the
spirits returning to the island are higher (cf. first *tanageru* myth,
unit N). However, the return of the novice is also the return of
the child who has forgotten everything he had remembered.
The Ongees regard this as an outcome of the body processing,
whereby all that is within the body is the blood and smell of
the pig and the words told to him by the chief initiator. Once
the spirits have heard the novice, the novice will be providing
them with whatever they want. This makes the novice, after his
return to the Ongees, a body that does not remember how to
'hunt and make food and tools for himself and others with
whom he lives as an Ongee—also he forgets all the *geyenene*
(weight) that he has to give to Ongees and thereby becomes a
recipient of *geyenene* from others. In this give and take the hide
and seek with the spirits becomes possible and so is the birth
of children' (Totanage, 1983). In the giving and taking of
geyenene (weight) the significance of exchange as a value and
virtue is highlighted within the Ongee culture.

Given this background and belief pertaining to the initiate
who has returned from the spirits' home, the next stage of
'ritual construction' (*gikonetorroka tamale*) is to give *inachekame-
ma*, (something that should not be forgotten). The Ongees now
come together to 'fill-in' the novice with ideas that should not

be forgotten. It is the retention of these 'ideas that makes the Ongee man a full and complete person in the social framework of expectations. Filling-in the initiate with memory is called *geyakwabe*, which in the Ongee language is a verb for 'transfer' or shifting something from one point or place to another point or place. In the context of *tanageru*, it is the 'memory' that is being transferred, being placed in the initiate. As the Ongees described it, the idea becomes clear: 'In *geyakwabe* we fill in and place the *inachekame-ma* in the empty initiate!'

Until now the *tanageru* dealt with the processing of the body. However, the stage and concept of *geyakwabe* adds another dimension to the ritual. It is the implanting of *inachekame-ma* within the empty body that makes the second phase of the ritual also a process of 're-socialization'. It is resocialization and not socialization, since the boy growing up within the community does learn about what he has to do and what is expected of him. However, at the start of *tanageru* he is processed into a child and is expected to remember only *gilemame*. Once the initiate reaches the home of Dare he is to convey what he has to remember as instructed by the initiator. In this going and coming back of the initiate, the only 'memory' he takes along with him is *gilemame*, which is a set of information establishing the give-and-take relationship between the spirit Dare and the visiting initiate. It is this memorized *gilemame* that not only enables the specific initiate and the spirit to relate to each other, but also makes the initiate capable of coming back to the Ongees. But the initiate who has returned went as a child and has come much closer to the spirits, since he has promised to give *gilemame* to them. The initiate who returns is not only like a spirit from the Ongees' point of view, but is also regarded to be an entity that is 'empty'. Thus emptiness is correlated with the state of not being heavy in the Ongee categories of thought. In the process all that the initiate had learned while growing up is forgotten and the community now has to tell him again about it. After coming back from the home of Dare the initiate is a person who knows about the spirits and his relation with them (especially through *gilemame*). What he has to be retold is the aspect of why he had to go up so that the seasons may continue, the spirits may descend, and basic life may continue. His part spirit personality has to be overlapped and fused with his identity of being an Ongee, with a human identity. Thus the Ongees have the *geyakwabe*, in which the novice may continue to be a person related to the community of the Ongees and not only to that of the spirits. Again, the structure of the *tanageru* ritual and of Ongee culture reflects the prime concern with the domains of *chongojebe* and *gengeyebe*, relations of give and take between humans and spirits and among humans. In fact, remembering all this constitutes a basic knowledge, which provides a technique for living with the spirits and forms an 'object' that gets 'transacted' in the ritual of *tanageru*, first, for the relationship between the initiate and the spirit, and then for the initiate to live as a novice within the society of his fellow Ongees.

The memory-implanting event of *geyakwabe* deals with knowledge about hunting, *enatandemetakwabe* (quick release of smell), which forms the founding principle of hunting, gathering, and living within Ongee culture.

Geyakwabe: *Event and Act of Implanting Memory*

Only after the dance of *yulelale-lale* is over and the initiate's body has been reasonably cooled, can the process of *geyakwabe* start. The skull of the pig hunted by the initiate, with cane woven onto it, is brought out by the chief initiator. The initiate's father hands over to his son a bunch of arrows made of bamboo splinters, and a small bow made of the rib bone of the pig hunted by the initiate. The small bow and arrow, about four to five inches long, are referred to with terms that are distinct from the term used for bow and arrow in the everyday context. The miniature bow is called *gekwa*, meaning one that takes the smell, whereas the general term for bow is *aa*. The miniature arrows, instead of being referred as *chenekwa*, are referred to as *ba?etuge*, which means 'going to the horizon'.

The chief initiator, holding the pig skull, starts walking around the initiate. It is a stylized walk where the chief initiator is bent forward on the ground. He holds the pig skull in front of his face and grunts dramatically, replicating the noise made by the pig. The chief initiator thus becomes the 'pig' for the people witnessing this event. The chief initiator's act is acknowledged by the onlookers when he scratches the ground just as

pigs do while searching for tubers in the forest. Every time the
initiator scratches the ground with the tusk of the pig skull, the
onlookers throw tubers in front of him. The initiator picks up
the tubers and places them in the basket hanging around his
chest. Every time this happens the pig's grunt becomes louder
and the onlookers are charged at by the initiator, and the circle
of onlookers scatters temporarily and one hears the shout,
'*edankuttuga tambanua*!' meaning a big fat pig.

After a few rounds of this play, the initiator holds the pig
skull in front of the initiate. The initiate, with the assistance of
his father, then releases the *ba?etuge*, small wooden arrow, by
means of the *gekwa*, small bow made out of the rib bone. The
arrow is released into the nostril cavity of the pig skull. As the
arrow goes through, the initiator gives out a loud cry identical
to that of a pig when it is killed in the forest. At this point the
father tells his son going through the ritual,

> Release the smell of pigs (*enatandemetakwabe*) again and again—and
> quickly and with care! then it will be the turtle whose smell you will
> release—you then will hide yourself for the honey and then you
> will hunt the snake, doves and nautilus. One has to kill them all,
> you have done it—don't forget it because you have to help others
> to do the same. Your killing will cause the winds to come, take away
> smells, and make all of us heavy. Do not be light and afraid. You
> have to be a hunter hunting the smells.

As this declaration ends, the chief initiator gets a nautilus
shell (*naratakwange*). According to the myth of the first *tanageru*,
it is in the nautilus shell that the 'snake' and 'bird' travel to the
home of the spirits (first myth of *tanageru*, units B, D, and E).
The chief initiator, holding the nautilus shell in his outstretched
hands above his head, goes around the novice imitating the
flight pattern of the mythical shell carrying the bird and the
snake. The chief initiator first replicates the movement of the
shell on the sea waves and then its flight up into the sky. After
a few rounds, the chief initiator holds the nautilus shell in front
of the novice. The novice, with assistance from his father,
releases an arrow inside the nautilus shell. The chief initiator,
imitating the dying shell, gives out a loud cry, and in an act of
struggle goes and lies down among the circle of onlookers. At
this point the father tells his son:

We have been killing dove, snake, and nautilus shell—one has to—then only we can hunt pig–turtle and get honey. You have done it. Now release smell of pigs and build your smell. In your heaviness is the increase in weight for all of us.

After this the chief initiator, who was all along being referred to as *muteejeye*, comes to the novice and declares:

I am your *mutarandee*, *muteejeye* has just left![1]

This marks the end of the chief initiator playing the role of the initiator *vis-à-vis* the initiate. The *muteejeye*, the chief initiator, now plays the role of *mutarandee*, which was his initial relation with the novice before the start of *tanageru*. This change from the *muteejeye* to the *mutarandee* is very significant within the progression of the *tanageru* ritual, especially in relation to the aspect that, through the *tanageru*, the individual male body is transformed so that it can visit the spirits. Once the individual returns, he has the identity of a child, and the child has to be 'resocialized' so that he gets incorporated into human society. This is why, with the change in the perception and function of the initiate, the initiator too changes into a *mutarandee*, an individual who is the prime socializer of the child, appointed even before the birth of the child.

In the act of shooting an arrow into the nostril of the pig skull and thus representing the pig hunt, the enactment also projects the killing of the '*muteejeye*' as an identity *vis-à-vis* the initiate. This becomes more explicit in the verbal communication declaration made by the initiator to the initiate. In the whole process involving the use of the pig skull, nautilus shell, various gestures, the enactment of hunting, and the verbal declarations, the Ongee community implants 'memory of hunting' in the initiate who has 'returned from Dare's home', and this becomes a possible goal attained in the *tanageru* ritual.

It must be noted that the shooting of the nautilus shell is an act that not only transplants memory for the initiate but also recreates the mythical event (myth of first *tanageru*, units D and E). The actual hunting of a pig by the initiate and the 'hunting' that took place within the first myth of *tanageru* are brought

[1] Saying 'has just left' is, in Ongee culture, a polite way of communicating the death of an individual.

together for the initiate in the ritual. It is the outcome of this fusion—forms of hunting, change of identity, and events happening in myth and in the actual life of the novice—that we can find a value created in the ritual of *tanageru*. When the initiate goes to the spirits and returns from the spirits' home, his journey brings together 'conventional value' and 'intentional value'. The belief that in every seasonal cycle an initiate has to be 'processed' and sent to the home of Dare is 'conventional value'. The necessity for the initiate to remember certain things and return from Dare's home, thus causing the start of the spirit-given seasons, is the 'intentional value'. On the successful return of the initiate, his identity with the spirits has to be broken down and his affiliation with the community of the Ongees who had sent him to the spirits has to be re-established. This is the task of *geyakwabe*, 'filling-in memory'. This 'filling-in memory' deals with aspects of knowledge about past events that are symbolized in the hunting of pigs and nautilus shells. It also deals with certain techniques pertaining to hunting animals without getting hunted by the spirits, which have implications for future events.[2]

In the *tanageru* ritual conventional value and intentional value being fused also brings together inter-subjective interests (as of men and spirits) with subjective interest (as of an Ongee person). Shooting arrows through the nostrils of pig skulls has implications for present Ongee hunters and gatherers as well as future ones. Shooting at the pig skull is the presentation of the 'symbolic sense' of hunting by smell. Shooting at the chambered nautilus shell, which is also enacted along with shooting at the pig skull, presents the novice with a 'symbolic reference' to the mythical event of the past. By bringing together the two values, two interests and two symbolic systems into one particular stage of the *tanageru* ritual, the initiate's identity is detached from the spirits' and attached to the Ongee community. This actually completes the process of ritual dealing with an initiate and making the individual initiate into a novice.

---

[2] Being aware of the past, i.e. spirits can hunt, is a knowledge which affects the present state of existence for the Ongee hunter and his 'technique' of living and hunting is in response to the knowledge of the use of clay paints, fire/smoke, and bones as forms of techniques which make it possible for the Ongee to hunt without getting hunted.

This change in the particular status of the initiate to novice is exemplified in the way the Ongees refer to the initiate. Upto the point of 'implanting memory' into the individual the individual is referred to as *naratakwange* (a chambered nautilus shell). After the individual has gone through the shooting of the arrow through the pig skull and the nautilus shell, he is called an *ale*, literally a child.

The *ale* (novice) is now taken by all the men to the coast. This procession is called *enondayabotakwe*, which literally means keeping the forgotten and the remembered together. *Enonda* means forgotten and *botakwe* means remembered. The two are joined by the verb *yabe*, which means keeping something safe and secure. The procession is led by the novice's father, father-in-law, and the *mutarandee* (who was the *muteejeye*). While the group of men walk to the coast, the men point out various directions and locations to the novice, telling him about the forests, the resources, and wind directions.

At the coast the novice is held up parallel to the water-line. For short intervals the men standing in knee-high water submerge the novice in the water. This sequence of dipping, *gegamebe*, is believed to be a means of keeping the memory in a child 'safely and securely' (*yabe*). The body submerged in sea water is seen as a way of cooling the body. It is this cooling that the Ongees believe causes the body to become heavy and retain memory. Indeed, the body of the novice, through *gegamebe*, is regarded as demonstrating and indexing its weight by showing no capacity to float. Floating is a quality attributed to the nautilus shell, but not to a child. *Gegamebe* is therefore an enactment and demonstration of the difference between the object that floats (nautilus shell) and the object that does not (the novice or child). This creation of a distinction through the sequence of dips in water is a reification for the Ongee that the novice's body is different. Unlike the initiate's body, which is processed and transformed to become like the nautilus shell (one that floats), his body is now cool and cannot float. As the men move out of the sea-water they tell the novice:

You [novice] child know the forest, you also know the sea—now in the forest and in the sea you can release smell while hunting—but if your smell is released, then with the winds the spirits will come

to take you away—remember that and do not forget that you could
go away with the spirits!

The *tanageru*, for the Ongees, is not just a ritual, but is an
essential and necessary event, process in which the continuous
and reciprocal movement between *chongojebe* and *gengeyebe* is
unfolded. The men having an impact on the spirits (*gengeyebe*)
and the spirits having an impact on men (*chongojebe*) are both
brought together significantly to create the distinction at the
level of the initiate and the novice, the one who visits the home
of the spirits and one who returns to the human world. This is
represented throughout the ritual by the community processing the initiate's body to send him to the spirits. Prior to doing
this, the winds are stopped—the spirits are offended and made
to move—and all this forms the *gengeyebe* domain. Once the
initiate returns he is a novice who is reminded of his being
subject to the spirits who cause death. Within the context of
Ongee culture, the *tanageru* ritual brings together the *chongojebe*
and *gengeyebe*, unfolding a reciprocal movement between
'practice of structure' and the 'structure of practice' (Sahlins,
1980) (*chongojebe* is practice of structure and *gengeyebe* is structure of practice). Although 'practice of structure' and 'structure
of practice' are brought together within the context of the
tanageru ritual, they are also created as distinct domains before
and after the ritual, in terms of the man-made seasons and the
spirit-given seasons.

By the time the men return with the novice, the *etotekwakabe*
around the novice's sleeping platform is completed. *Etotekwakabe* refers to a cage-like construction made of sticks. Sticks
that are about five feet high are tied at intervals of about four
inches around the sleeping platform. The enclosure around the
sleeping platform is covered with sheets of cloth to protect the
novice. When the novice returns the winds and the spirits start
visiting the island. The enclosure is a way of keeping the spirits
away from the novice. *Etotekwakabe* literally means keeping no
wind and no smell.

Once the novice enters the *etotekwakabe*, he cannot be seen
by outsiders. Things which may be required by the novice, such
as his water containers, smoking pipes, and sleeping mat are
all placed in the *etotekwakabe* before he enters the enclosure,

including a good supply of betel-nut and leaves. The novice is accompanied only by those men who have been through the *tanageru* ritual. Once all of them are inside, the novice is served with his first solid food during the *tanageru* ritual. This food, *ekwaikata*, comprises roasted jack-fruit seed that had been collected and processed (through sun drying) during the honey-gathering season.

Before the serving of *ekwaikata* the initiate was served only the blood of the pig he had killed. In relation to this, the serving of *ekwaikata* marks the novice as *ale*, a human child who is not only heavy but also distinct from the spirits, yet who cannot consume solid food. After the feast, one by one, all the pandanus leaf ornaments (*owetorakabe*) are removed from the novice's body and passed on to people waiting outside the sleeping platform, who in turn hang the *owetorakabe* on the 'leaf curtain'. Once the novice's body is devoid of any adornments the process of *utan-name* (birthing) starts. In the course of the *utan-name* the men inside the enclosure adorn the novice with a different set of ornaments, which mark his capacity to handle weight. The material of the ornaments now used are not leaves and strings but relatively heavier things, such as glass beads and strings of *chendaru* (tusk shell).[3] Every family at the camp contributes in making and giving the somewhat heavier ornaments of glass beads and shells. These ornamental shells are collectively referred to as *enebule*. The body of the novice is literally 'loaded over' with *enebule* and with this adornment the restriction on the novice of keeping his eyes closed is lifted, and he can hold his head up.

Now the *gebule*, the cold state of novice's body attained by previous steps of ritual, such as *yulelale-lale*, is partially heated up by the application of red clay paint. This clay paint is mixed with the fat of the pigs hunted by the initiate. As the body of the novice is painted in glossy red paint and *enebule* are placed on it, the process is completed by the ceremony of *getulakwoyebe*. *Getulakwoyebe* entails giving string ornaments to the novice. These string ornaments, *tobegeni* and *ugetamabo*, were made by the women during the season of Torale. *Getulakwoyebe* is yet another phase of *tanageru* ritual by which the community

---

[3] *See* Radcliffe-Brown, 1964, pages 477-80, for strings of tusk-shell necklace.

has processed and completed the construction (*tamale*) of the 'child' who has returned from the home of the spirits. The spirits, the winds, and the spirit-given seasons are yet to come.

Onotandeme: *Duration of Continued Pig Hunting*

In the middle of June the weather conditions change. Occasional downpours, strong gusts of wind, fallen trees along the coastline, and a turbulent sea, all mark the strong south-east winds. These winds, ranging from 55 to 60 km per hour, reshape the coastal area with their rage. With the rising temperature, day-to-day storm conditions develop and the wind blowing from south-east starts moving south-west. Under these changing meteorological conditions, the slowly changing novice and the other men continue to hunt pigs, day after day. The hunting of pigs by the novice under changing wind conditions is called *onotandeme*. As the days go by the scaffold and poles meant for hanging the pig skulls get filled with neat rows of pig skulls.

Every time the novice returns successfully from a pig hunt, he is made to sit on the pig and feed on pig blood. However, no massage is given to the novice after the blood has been consumed. As the days go by, large quantities of pig meat have to be managed, and therefore the cooking area, *kwotaya*, keeps expanding with extensions of thatch and fireplace. (In the day-to-day context, the cooking area is referred to as *kalakale*, *kwotaya* is a term reserved for the cooking area only during the rituals.) Since not a single portion of meat may be wasted, and there is more meat in supply than the level of consumption, leftover burnt and decaying meat is packed in large baskets and kept hanging over the cooking fire. This odorous meat, *nanguchumemy?*, fills the air with its distinctive flavour at the camp-site. The Ongees, generally very concerned about the smell of decaying substances, reflect a very positive attitude to *nanguchumemy?*.

The shifting wind conditions, the continued pig hunting, and the excess of pig meat that starts decaying and stinking (*nanguchumemy?*), mark the whole ethos of the camp-site with the novice's accomplishment and characterize the duration of *onotandeme.* The word *onotandeme* is significant in its own right, especially in light of the novice's accomplishments. The word *onotandeme* is composed of the prefix *onotota* (responsibility) and suffix *ndeme* (bearing or taking), which means responsibility bearing or responsibility taking.

Around the first week of July the scaffold constructed for hanging the skulls of hunted pigs was nearly overloaded, and the camp was talking about bringing an end to the *tanageru* ritual. The shift in wind direction was experienced by everyone; it was expressed as:

> Pigs will soon become lean — all the tubers are gone, we shall end the ritual!

On the evening when the last pig skull is placed on the scaffold, a sing-song session called *eneyenjulebe* is held. All the outer covering from the sleeping enclosure of the novice are removed. The men go inside the enclosure and congregate around the sleeping platform for the singing session. The chief initiator leads the singing. This special singing session should be more appropriately described as a session of humming and shouting in chorus. Everyone joins the chief initiator in chanting *'galee me hmmm!'* meaning 'it is nearby!' After a few repetitions the uncoordinated chanting develops into a high-pitched, well-harmonized crescendo. The last part, *'hmmm'*, especially has its own impact, when all the men breathe out vigorously. *'Hmmm'* is also the part in the whole recitation in which the novice participates by repeating it with others. It is also at this point that all the men surrounding the novice apply a little pressure on his body, and in that process bend slightly forward. This leads to a constant shifting and shuffling of bodies around the novice. As soon as a little space is available, one more person enters to sit in the enclosure. As the singing proceeds the enclosure becomes more and more crowded.

Every time the synchronized crescendo builds up and the people have adjusted around the novice, the chief initiator breaks the harmony with a loud shout saying:

> Die die quickly you nautilus shell you are here a child back with us—you have finished your work—now you can speak to us—you have been brave for yourself and for all of us—we will tell you soon

about all the place around us and what happened in it while you were away at Dare's home.

As this declaration by the chief initiator ends, all the individuals shout, turning their faces upwards. Each individual speaks about a different subject, but all the individuals speak at the same time. It is at this particular moment that the individuals shout about which household had a new child, which house is expecting a child, where death occurred, who has been very successful in hunting different animals, who got married, who wants to get married, who made the canoes, how much honey was collected, and who paid visits to whom. The information given to the novice is not only about events during the period of his visit to Dare but more than that. What the novice hears about is actually all the things that happened during the entire seasonal cycle.

Early in the morning, after the night-long singing session, the chief initiator and his wife take all the bones of the pigs hunted by the novice and bring them to the mangrove forest. The novice's parents dismantle the enclosure around the novice's sleeping platform. The extended sleeping platform is now divided up for each family to occupy. However, the thatch overhead remains undivided. This reorganization re-establishes every family's identity in the camp, in terms of having its own individualized *kame*, sleeping platform, and starts the ritual of *onebelatekwebe*.

Onebelatekwebe: *A Step Towards the Completion of the* Tanageru *Ritual*

Onebelatekwebe is held after the bones left from the pigs killed by the novice have been buried, and every family has its own *kame*, separated and distinguished from those of his neighbours.

The novice is called out after all the preparations are completed, and for the first time he emerges from his sleeping quarters without any assistance. As he comes out he proceeds to his chief initiator's *kame*, and when he reaches it, the chief initiator's wife says,

See—See well, search for it—there is nothing we did not do anything wrong—the winds are here!

On hearing this, the novice very gently does *kwayaya*, which literally means 'Where is it?' However, in the context of the ritual, *kwayaya* also entails a dramatized act in which the novice overturns several water containers, overturns baskets, unties some of the bindings of the *kame* and *korale*, and scatters a few things around. This scattering, spilling, and throwing around is a display of anger. However, this dramatic display of anger is made into an occasion of 'grab all you can' by the children standing right behind the novice. As the novice succeeds in displaying his anger at the chief initiator's *kame* and the young children grab whatever has been thrown around by him, the chief initiator's wife shouts:

Kuge, Kuge, Kuge! go see elsewhere!

The chief initiator then directs the novice to the adjacent *kame*. At each *kame* the novice goes through the acts of *kwayaya* and *kuge*. *Kuge* in the Ongee language means war, anger, and stone. As the novice finishes at one *kame*, people in the neighbouring *kame* put out things that he can spill and scatter. The Ongees explain the *kwayaya* as a realization by the novice that Dare had told him not to hunt pigs, which of course have been hunted. So, on behalf of Dare, the novice wants to find where the bones of the slain pigs are, the idea being that once the novice finds the bones he can identify the consumers of the pigs, which were reserved for Dare to consume. Ongees are aware that after *eneyenjulebe* is held the novice will speak and express himself. However, the novice also has lost his memory, especially of his own hunting of the pigs and responsibility for them to the camp. This 'manipulation' and 'transplantation' of memory is achieved by the *ge?kwabe* performance within the *tanageru* ceremony. Because of the prior two stated, *eneyenjulebe* and *ge?kwabe*, the chief initiator and his wife collect all the leftover bones of pigs consumed and hide them. Since all the bones have been buried (referred to as *eleborale*), the novice can now express but lacks memory about he himself hunting it. *Kwayaya* and *kuge* are perceived by Ongees as natural 'childlike' behaviour on the novice's part. Ongees do not appreciate or approve of people getting angry and expressing it either in silence or destruction of things. To get angry is a quality and behaviour that Ongees associate mainly with spirits and children. Often when children in the camp wanted to express their frustration, dissatisfaction, or register protest they could throw shells and chunks of corral in the centre of the camp-ground. This would draw all the camp-mates to the child to pacify him by giving him whatever he wanted.

Power Implications, Expression of Anger, and Relation of Man and Spirit

The novice's expression of anger at the camp-site coincides with the anger expressed by the spirit Dare along the coastline in the middle of July. Both forms of anger are significantly related, in the Ongee point of view. In the words of Teelai at Dugong Creek:

> Dare is angry—Dare is about to leave— *tanageru* is over—all the pigs have been hunted—*ale* (novice) knows that angry Dare is going so he looks for the pigs of Dare which Dare could not hunt! Dare and *ale* both are angry, soon Dare and *ale* both will lose anger. *Ale* will remain with us—Dare will leave us!

From the Ongee point of view, the destruction caused by the strong south-east winds in the middle of July indicates the departure of the spirit Dare. Dare leaves, and in that process the novice is also left behind by Dare. Dare brings the novice back, since the novice had succeeded in convincing Dare that he (the novice), along with the other Ongees, would make provisions for all the spirits to consume specific food items each season. However, when Dare arrives on the island, Dare finds no pigs. All the pigs have been hunted by the community during the *tanageru* ceremony. It is here that the Ongee culture's logic pertaining to the human and spirit relationship is revealed.

Since Dare comes with the novice to eat pigs, the human community 'gets back' the novice. In the process of sending the novice up to Dare, not only is the novice's body processed, but a large number of pigs are killed. It is this killing of pigs that is 'offensive' to the spirit Dare. On not finding what Dare came down for, Dare departs. Thus the Ongees cause Dare to depart by killing the pigs which Dare was meant to hunt. However, by doing this, the Ongees cause the next season to start, with the winds changing from south-east to south-west, marking the arrival of Kwalakangne?. This is referred to as *kwayaya*.

Within the sequence and progression of *tanageru*, *kwayaya* represents a significant juncture. *Kwayaya* is the point where the Ongees and the spirits really enter into a relationship, in which the humans succeed in affecting the spirits associated with seasons and winds. *Kwayaya* marks the following gains of the human community from the spirits.

1. The Ongees get back the individual they had sent to the spirits by processing his body. In all other circumstances the Ongees take precautions so that they may not end up going to the spirits' home.
2. The spirits are responsible for birth. It is the spirits who come and get transformed through the process of being trapped in food substances and, consequently, in women's wombs. However, *kwayaya* marks the point where the Ongees cause the birth of an individual (the initiate / novice), by processing the body and creating a situation in which Dare leaves in anger on not finding pigs, but leaves behind the initiate.

During *tanageru*, human beings cause 'birth' and 'death' that are generally the responsibility of the spirits. This is endorsed, in that at the juncture of *kwayaya* the novice behaves exactly like the spirit Dare, in terms of destruction and the expression of anger at the camp-ground. (Dare's anger is visible in the coastal areas.) The men who cause the departure and arrival of the novice by means of *tanageru* become identical with the spirits who, in taking away a person, cause death and, in sending him back, cause birth. The time preceding the ritual of *tanageru* and the time after the ritual are thus framed by offences, 'bad-work', in which the Ongees consume food substances that are reserved for the spirits and are associated with specific seasons. By doing this offensive work the Ongees create a duration in which the spirits and the winds are absent. Once the spirits and winds are gone (over which men generally have no control), the Ongees perform the *tanageru*. So in the absence of the spirits that has been caused by Ongee actions, the Ongees do what the spirits do, that is, they enact the 'symbolic death and birth' of the initiate. And to get past this phase, to bring

back the seasons and the spirits—one more offence—the Ongees enact the *kwayaya*, when the novice and the spirit of Dare
get angry over not finding pigs. The spirit leaves and then
comes Kwalakangne?.

This makes the structural pattern of the *tanageru* an initiation ceremony homologous to that of the 'domains' mentioned
earlier. That is to say, both during *tanageru* and at other times
in the seasonal cycle, the Ongees and the spirits have mutually
inverse capacities for 'causing' birth and death or presence on
the island. Consequently, the relation of power between spirit
and man in the Ongee world-view is distributed so that both
man and spirit have similar acts and impacts on each other, as
embodied in the *chongojebe* and *gengeyebe* domains.

However, although the distribution of power between man
and spirit may be equitable in terms of 'acts' and 'impacts', it
is also 'asymmetrical' and 'self-reciprocating'. This asymmetry
occurs because within the space shared by the Ongees and the
spirits, both have to hunt and gather, and they both cannot hunt
and gather the same food at the same place at any given point
of time within the seasonal cycle. Therefore the Ongees often
say,

> All food is around us through the seasonal cycle but we should not
> hunt what spirits are eating in a particular season. If we do that,
> spirits will get angry and leave us, then no winds will come—
> *igagame* will not change. There will be no childbirth. Long [in
> duration] anger of spirits is not good.

Thus the Ongees create the absence of spirits and winds for
holding the *tanageru* and end it by causing the presence of
Kwalakangne?.

Last Phase of Tanageru: *Pacifying and Incorporating the Novice*

After the spilling and throwing of things at each *kame* in the
camp, the novice returns to his own sleeping platform. The
chief initiator then requests the 'angry' novice to step out. The
novice, after displaying a certain amount of resistance, steps
out. Both initiated novice and the initiator then start visiting
each *kame* for the second time. This time the visit to each *kame*
has a different purpose, and consequently involves a different
enactment.

In front of each *kame*, the head male of the family and the chief initiator stand on the two sides of the novice, and pull him down on the *kame*. All three bodies land on the *kame* with a loud noise and, for a moment, the women sigh with concern. As all three bodies land on the *kame*, the novice is tickled by the men on both sides. This tickling is called *geregerabe*, which is literally glossed as 'putting on weight'. The tickling continues till the novice breaks his silence and cannot resist laughing.

When the novice loses his self-control and breaks out laughing, the men help him to get up and the chief initiator takes him to the next *kame*. Again the tickling is done and one by one all *kame* at the camp-site are visited. This visit to each *kame* associated with an individual family by the initiate is called *tuleye?keneyabe*.

The act of tickling is regarded by the Ongees as a way of making a 'child' or a 'person' happy. Pulling down the novice on a *kame*, from the Ongee point of view, is a demonstration of the novice's heaviness of body, which falls down and topples with little effort. The Ongees believe that making someone break his silence and laugh is a way to end and pacify anger.

When the visits to each *kame* at the camp are over, the novice approaches the centre of the camp-ground, cuts apart the leaf curtain and sits down. It is evident from the novice's face that he is in a state of pain as a result of having been pulled down at each *kame* and being tickled violently. As the novice sits in the centre of the camp-ground, his wife, mother, and mother-in-law come to sympathize and say reassuringly,

Do not be angry! not any more—you will always be laughing with us! *Tanageru* is about to end be with us—do not be angry, we all will make you happy—be heavy!

The novice shakes his head in silence and with seriousness to confirm that he has understood the women's point of view. As the women leave the novice, the chief initiator's wife with a large new *toley* (basket) comes to the centre of the camp-ground. The basket is kept in front of the novice. In this the chief initiator's family puts various kinds of *maonole* (tools), which include newly-made knives, arrowheads, adze, and sharpening stones. This starts the sequence of *mandy?eonebe*, gift-giving, during which all the households come together and

give something to the novice. Apart from *maonole*, the camp
gives small baskets, arrow shafts, small canoes, sleeping mats,
strips of cloth, and dogs. Through the series of gifts the novice
acquires all that an Ongee man needs to function as a hunter
and a gatherer on the island.

The Ongee idea of giving gifts, *mandy?eonebe*, to the novice
is intended not only to make him happy, but is also seen as
loading him excessively with things so that he becomes heavy
and is compelled to stay on the island and 'use the gifts' to live
within the human community. The Ongees also believe that if
gifts are given then gifts can be received, which is an important
way in which relations are established and managed. The
relations of gift-giving and receiving, known as *galematabe*, is
such a central and significant idea, that the Ongees identify the
spirits and the other tribes of the Andaman Islands as all those
with whom *galematabe* is done. This also distinguishes the
non-tribal population and its spirits as people with whom they
have no *galematabe*.

As the novice accumulates his gifts, he puts them all in two
or three baskets and hangs them from his forehead, thrown
over his back. With this load, he visits each *kame* where he was
tickled and each household from where he received a gift. At
each *kame* the novice reaches over his shoulders without looking and pulls out whatever he can get hold of. This randomly
picked object is then given to the head of the household being
visited. The outcome of this process is that every household in
the camp gives gifts to each household. This giving and receiving of gifts, as exhibited in the last phase of *tanageru*, reifies the
significance that the Ongees place on give-and-take relations,
a characteristic relation by means of which the Ongees relate
among themselves and with their spirits.[4]

The following morning all the women gather and go to the
creek. The month of June is now about to end, and creeks are
filled with rainwater and teeming with various kinds of crabs.
It is to collect crabs that the women go to the creek. On their
return in the evening, a *gakuwejebe*, a crab feast, is held for all
the members of the camp except the young boys. (A young boy
can eat crab meat only after he has gone through the *tanageru*

---

[4] However, the give and take relations *vis-à-vis* spirits are referred to as *gae?bebe*.

ceremony.) On the evening of 27 June 1984, the *gakuwejebe* was
held for the two novices who were initiated in the course of my
field-work. By this time the two novices had killed 72 pigs, and
before the *tanageru*'s last procedure, the feast of crab meat was
to be held for them, in which they would be served solid crab
meat for the first time. While the women are away collecting
crabs for the feast, the men start dismantling all the *kame* and
thatched shelters. Before the sun sets, the camp will shift from
the forest to the coast, and it will be in this new location that
the crab feast will be held.

Apart from the novices and the young unmarried, uninitiated boys, all the males of the camp help in shifting and
setting up the new camp-site at the coastal area. The new camp
at the coast, reconstructed out of the old material, is called
kogeye?be. In setting up this camp all the families set up their
kame in a circular manner and each *kame* has its own roof. While
all the men do the work, the novice and his young friends sit
along the shore and make *gerugerube*, wind-wheels. The wind-
wheels are made by folding two strips of dried pandanus
leaves criss-cross and then sticking them on a twig. The novice
makes the *gerugerube*, and all the children hold them up in the
wind and run around the new camp-site shouting '*tototey!*
tototey!' (winds! winds!). As the children run around the
gerugerube start spinning in the slight wind from the seaside.

By evening the camp in the coastal area has settled down.
The women return with loads of crabs. At night the camp has
the meat of boiled crab. All the scraps are carefully collected so
that the following day they can be thrown into the creek. The
crab shells from the novice's quarters, however, are retained
along with the pig skulls procured in the course of the *tanageru*
pig hunting. The novice's mother collects all the crab claws,
which she dries to make smoking pipes. At the end of *tanageru*,
once the crab claw smoking pipes are ready, they are distributed among all the Ongees living on the island. The following day, early in the morning after the night of the crab feast,
the novice, the initiator, and few other men leave the camp for
the final pig hunt with which the *tanageru* will end. This final
pig hunt is different in many respects from the earlier ones in
the course of *tanageru*.

On reaching the forest, the novice is left alone to hunt down a full grown boar. In all previous hunting expeditions, within the duration of *tanageru*, the initiator accompanies the initiate in the step-by-step killing of the pig. Unless the novice hunts down a boar entirely on his own he may not return to the camp, and only after that can *embuteye*, the last phase of *tanageru*, be undertaken. On successfully hunting the boar, the novice returns to the coastal camp. This is distinct from the previous returns of the novice after hunting. No tree buttresses are beaten, since the novice returning from the forest is now regarded as an Ongee coming back, and not as a person who is between the worlds of humans and spirits. It is no longer as the dangerous and ambivalent person returning from the forest that the novice carries the load of the hunted pig on his back, demonstrating that his body is no longer light, but strong and heavy enough to carry the weight.

On reaching the camp-ground, the novice stands in the centre of the ground. His father and mother in their *korale* embrace each other. Holding each other, *enegeteebe*, they cry solemnly. The chief initiator and his wife visit the novice's parents. The initiator's wife then consoles the novice's parents by saying,

My husband [initiator] has brought back the child. Do not cry. He is no more a child of yours, he is ours too! He now belongs to all of us!

As the initiator's wife is consoling the novice's parents, the initiator transfers the load of the hunted pig from the back of the novice to the front. Now the pig hangs, tied up in bark strips, from the novice's neck. All the camp-mates congregate around him and one can hear statements like— 'Our child has become strong and brave', 'He has learned the skills!', and 'He stands firm with all that weight'. People at the camp admire the completion and success of the novice's *eneyalelabe* (learning).

Now the ceremony of *embuteye* (literally, to strip) starts. Men remove the load of the hunted pig from the novice's body. The boar is placed in front of him. The novice is then stripped of all the body ornaments and the *atakee* (the strip of bark tied around the waist) is removed. Finally, the novice stands stark naked, and all that was on his body is placed on the boar. Indeed, the ornaments and the *atakee* are carefully tied around the boar. The novice is now covered with a new *atakee* and the glass bead string is placed back on his body. *Embuteye* ends with 'dressing the pig' and 're-dressing the novice'. After the novice has been adorned in his new *atakee*, the novice's wife brings a basket-load of firewood and some chunks of smouldering fire and starts singeing the pig. As the whole of the pig gets charred, the novice cuts it up and divides it into two portions and sets it up in two different containers to cook.

This is the first time in the novice's life that he hunts, processes, cuts up, and cooks the pig all on his own. The Ongees regard this as not only giving an individual male the right to hunt and cook, but see it more as a demonstration of the novice's identity as an Ongee man, completely processed. *Eneyalelabe*, learning, is over. The novice can now do all that an Ongee man should do, i.e. hunt and feed.

In about three to four hours the meat is cooked. One out of the two portions (divided by him) is taken by the novice to his chief initiator's *korale*. These portions are called *gangeebe*. I was told by the novice that he had to give *gangeebe* to his chief initiator because he realized how much the initiator had taught him and how much help he had received. Above all, in the course of *tanageru*, the *muteejeye* (initiator) has not eaten any pig meat (just like the initiate), so it is a case of 'my asking him to eat pig again!' In other words, *gangeebe* is a token of 'thanks' in recognition of the assistance given by the initiator to the initiate.

Upon returning from the chief initiator's *korale*, the novice invites all the people to come and take a portion of the pig meat he has cooked. The novice's invitation is readily accepted. All the camp-mates sit in front of their respective *kame* facing the camp-ground. The novice is the last to help himself with what he has hunted and cooked. As the novice starts eating pig meat, which he refrained from during the course of *tanageru*, the entire camp shouts '*enangatokuwa*', meaning it is complete.

As the novice finishes his portion of meat, his wife brings a container of white clay. Like all the others and at all other times, the novice covers himself with white clay after consuming meat. This makes the completion of *embuteye* the marker of the incorporation of the novice into the Ongee community. This aspect is reflective of the completion of learning and transformation of the novice's body, whereby the novice hunts alone, brings the boar on his own, processes it, cooks it, and serves it to his camp-mates and, finally, consumes it himself.

The day after *embuteye* is like any other day in the life of the Ongee community. It is the season of Kwalakangne?, and the Ongee are settled down at their coastal camp-site. *Tanageru* is over. The end of *tanageru* is indicated by only one clue. There are scaffolds lined with the 72 skulls of the pigs killed in course of *tanageru*, and all have cane woven around them. After *tanageru*, every man saves the pig skulls of his *tanageru* hunt. As the time goes by and residence is shifted, some of the skulls are damaged or lost. However, one or two are always carefully perserved in each *korale*.

The community will now start hunting turtles. The novice too will accompany them, since it is only after the *tanageru* pig hunting, in which the novice goes through phases of lightness and heaviness homologous to the nautilus shell, that he can go to hunt turtles with harpoons. This is particularly important since 'to hunt turtle the hunter has to submerge himself in sea. If he is not capable of "floating" like a nautilus shell he is no good as a turtle and pig hunter'. Those have been initiated are allowed to actually hunt turtles. Uninitiated men accompany turtle hunting expeditions but they never do the actual hunting.

On asking Totanage's son, whose *tanageru* was held during the course of my field-work, he told me (after his *embuteye* was over),

> It is over, Dare is gone Kwalakangne? has come, I am a *gayekwabe* like all other Ongee men who have gone through *tanageru*. I am now *gengeyebe*.

Gayekwabe is the Ongee term describing the quality of a body because of which smell is released in relation to safety and danger. Indeed, the *gobolagnane* are used for the purpose of maintaining, releasing, and restricting an individual's body smell. It is the capacity of the body to function like a *gobolagnane* that makes it *gengeyebe* and complete and heavy.

CHAPTER 8

Conclusion

> There is a belief that mortals wandering by themselves in the jungle
> have been captured by the spirits. Should the captive show any fear,
> my informants said, the spirits would kill him, but if he were brave
> they would take him to their village, detaining him for a time, and
> then releasing him to return to his friends. A man to whom such an
> adventure has happened will be endowed for the rest of his life with
> power to perform magic. He will pay occasional visits to his friends
> the spirits. [Radcliffe-Brown, 1964: 138-9.]

The world in which the Ongee hunters and gatherers continue
to live and die is a cultural construct of nature, in which the
Ongees belong to a scheme of hierarchy and stratification, to
which both dead relatives, who have become spirits, and the
animals belong. In the context of this hierarchy of relations,
spirits, human beings, and animals share a common space
through which all of them move. The places constituting this
shared space are specifically marked by seasons, wind conditions, moisture, and the available food resources, all of which
are conditions determined by the presence or absence of spirits
in a given place. In each place, spirits, humans, and animals are
related to each other through the act of hunting, i.e. the act of
taking life. The spirits hunt and gather Ongees and animals; the
Ongees hunt animals; and both Ongees and animals try to
avoid being hunted.

The Ongee cultural construct of nature is based on an
intricate balance of power between the powerful and the
powerless. The aspects of power are best demonstrated and
accounted for by the Ongees in their activities of hunting and
gathering. The hunting of an animal embodies the success of
the hunter in taking away the life of an animal, but this life-
taking activity also contains the aspect of life generation, since
the prey provides the hunter with the capacity to live. In the act

of hunting the prey becomes powerless and the hunter asserts his *enakyu?la* (power) by successfully moving to where the animal is and by successfully keeping his own body smell from reaching the animal. The Ongees regard this act of hunting, involving aspects of movement and smell, as a *talabuka* ('conjunction') of the paths of movement between the powerless prey and the powerful hunter. A failure in the act of hunting is the failure of the hunter's capacity to move and to control the movement of his smell that reaches the animals. If the hunter's body smell reaches the animal before he reaches it, the animal avoids the conjunction by moving along a different path, and thereby acquires the power necessary to deal with the powerful hunter. In such a case, a shift occurs which the Ongees regard as a change from being powerless to being powerful.

In the Ongee world-view, the relationship between humans and animals is transposed onto the relationship between humans and spirits. The spirits have the power to hunt and gather the powerless Ongees. The smell released from the bodies of the Ongees can attract the spirits down to the island. The smell carried by the winds brings the spirits to the human body and the spirits take away the Ongee, resulting in the death of an individual within the community. The loss of an individual Ongee within the community, like the loss of a hunted animal, has another aspect to it. For the spirits who cause the death of the human individual, it is the addition of yet another spirit to their community. The Ongees taken away by the spirits, regarded as being 'hunted and gathered by the spirits', become spirits.

Human beings who are transformed into spirits through the process of being taken away by the spirits and through death are different from the living human beings on the island in the following ways:

Ongee Human Beings	Ongee Spirits
Have jawbones and teeth to masticate.	Do not have any capacity to masticate therefore depend on sucking nourishment or on entering the food substance itself.

Ongee Human Beings	Ongee Spirits
Have bones, which are the prime source of smell, giving a concrete form to the living being.	Do not have any bones, therefore are living things devoid of any smell and concrete form.
Due to their weight (from the bones) and form, they are capable of moving only on the horizontal axis.	Having no bones, the spirits are capable of moving anywhere and everywhere along with the winds, which makes them uniquely capable of moving on both the vertical and horizontal axes of movement.
Release smell and have the capacity to perform the operations of cutting, binding, and fabricating tools.	Absorb smell. Do not have the capacity to bind or tie things in order to make tools. However, they have the capacity to sharpen the cutting edges by grinding them on appropriate stones known only to the spirits.
Have the capacity to hear and see.	Do not have the capacity to hear or see. Depend on the capacity to smell and touch.
Move from place to place in relation to the movement of the spirits.	Move from place to place within the entire space along with winds, bringing about changes in places.

The spirits come down to the island with the winds in search of the Ongees and their tools. By taking an Ongee, the community of spirits gains the human capacity to fabricate tools, which they are not capable of doing on their own.

Spirits come to the island in search of food. Since spirits do not have lower jawbones, they lack the capacity to masticate. As a result, they prefer soft and liquid food which they consume by entering into the food substance. Honey, therefore, plays the most significant role in the Ongee diet and in their rituals and world-view.

Women who eat food substances such as honey, in which a spirit is trapped, become pregnant since the food, once consumed, releases the spirit who then becomes a foetus. Consequently, the coming down of the spirits to the island is a pattern of movement with which the Ongees associate childbirth, an addition to the community of humans, and the end of a spirit's. This coming down of the spirits on the vertical axis is diametrically opposed to the spirits going up on the vertical axis of movement. The upward movement of the spirits is associated with death and the transformation of human beings into spirits, whereas, the downward movement is associated with the birth of a child and the transformation of a spirit body into a human body.

The transformation of spirits into humans and of humans into spirits, as embodied in vertical movement, enables the spirits and the Ongees to share an identity but also means that they have different capacities, locations, and body conditions. The system of birth and death thus includes powerful spirits and relatively powerless humans, where the latter are subject to the power of the spirits in terms of death.

The spirits' movement down to the island in search of food and tools poses a constant threat of death for the Ongees. The Ongees deal with this possibility of death by using clay paints on their bodies, keeping fire and smoke around them, and preserving the bones of their dead ancestors; the Ongee term for these various objects and substances is *gobolagnane. Gobolagnane* restrict and control the smell released by the human body which attracts the spirits. The use of the *gobolagnane* on and around individual Ongee bodies, at the camp-site, and while moving across places makes it possible for the Ongees to move safely in relation to the spirits. In light of the differences and the concerns shared between spirits and humans, Radcliffe-Brown (1964: 307) rightly posited that within the Andamanese world-view men and spirits are opposed to each other. They are opposed to each other but they do belong to a single system of smell and relations of power which are formed on the principle of the movements of smell and bodies. Ongees and animals alone have bones which distinguishes them from the spirits who do not. Bones are the most hardened, heavy, condensed form of smell. The solidification of smell in various degrees forms liquids and fluids, flesh, and bones within the body. An increase in heat leads to the release of body fluids and, finally, to the disintegration of bones, leading to death or *nanguchumemy?* ('decay'), and the formation of a spirit body. Consequently, the spirits' bodies, devoid of bones, do not exude smell but only absorb it, and their bodies are soft and light in weight. The light weight of the spirits enables them to move any and everywhere with the winds. The potentiality of the release of smell from the living body and the absorption of the smell by the spirits sets up the power relation between the Ongees and the spirits, as a result of which all movements are affected and connected. Since spirits can move anywhere and everywhere with the winds, every human movement makes the use of *gobolagnane* a constant necessity. The use of *gobolagnane* makes it possible for the Ongees to move in relation to the spirits by limiting and controlling the movement of body smell.

Each *gobolagnane* manipulates smell in a different way. The use of white and red clay paints exemplifies Ongee efforts to alter their body conditions in terms of experienced heat. When the body experiences excessive heat then smell is released in the form of sweat. The Ongees believe that by retaining or releasing smell the weight of the body is affected. Cool white and hot red paints mark the various states of an Ongee body. The application of white clay paint cools the body, confines the smell, and keeps its weight heavy. If a body is heavy the spirits cannot lift it up and *metakabe*, 'fly off'. Conversely, the application of red clay paint makes a body hot and causes it to sweat, thus releasing both smell and weight. A sweating body attracts the spirits and, because it is light, the spirits are able to lift it away. Ancestral bones, which are the hardest and most concentrated form of smell, are kept by the Ongees in small baskets. When the Ongees wish to enlist the aid of their ancestral spirits, they uncover the bones and cause them to release smell which then attracts the ancestral spirits. The smoke of fire offers yet another means of protecting the Ongees from the spirits. The smoke from the fire acts as a sheath under which the Ongees can move safely from place to place since the smoke prevents the winds from picking up the smell of Ongee bodies and carrying it to the spirits.

Spirits also come to the island for Ongees because they alone have the capacity to make *maonole* ('tools'), i.e. the capacity to bind and cut things in order to fabricate tools. To avoid the loss of *maonole* and of life (the Ongee term for life is *teenehye?bagabeh*, which literally means to cut and tie), the Ongees use *gobolagnane*.

The Ongees perceive the distinction between humans and spirits in the following manner:

> Spirits can get food and from them we can get stones to sharpen our *tejage* ('cutting edges') children. The spirits are always here on the island in the forest and in the sea, they come to get us and take us away like turtles from the sea and pigs from the forest. We Ongees are afraid of the spirits, but we have a give and take relation with them. The spirits need us because only we can get food as well as make things like arrows, bows, adze, knives, nets, baskets, mats, ropes, pots and containers, and smoking pipes.

This giving and taking between man and spirits, that is, between death and birth, and between the movement of smell and of the body forms the core of reasoning which enables the Ongees to undertake seasonal translocations and, hence, avoid being in a place where the spirits are hunting and gathering.

In the world-view of the Ongee hunter and gatherer, where the power relations among men, spirits, and animals are set up through the basic metaphor of the hunter and the hunted, and where the whole pattern of movement and awareness of smells, winds, and spirits is structured in relation to Ongee movement, we see a picture which is akin to that of two individuals sitting at two ends of a see-saw. The powerful and powerless, whether human beings or spirits, are like the two individuals at the two ends of the see-saw. They both have an effect on each other, in terms of weight distribution and displacement, so that the motion of the see-saw can be continued. The Ongee notion of power, as exemplified by the patterns of movement and needs of the spirits, are like the efforts of the two individuals on the see-saw. The place where the spirits are is the place where the Ongees are not supposed to be, in order to prevent encounters which lead to *enegeteebe* ('death'). Just as the two people are essential for the see-saw to continue moving, the spirits and the Ongees are both essential to each other. Because of power, in terms of the movement of one in relation to the other's presence or absence, the life of the Ongees is constructed from reciprocal actions and reactive forces between the powerful and the powerless. It is the movement which is reciprocal and the *gobolagnane* embody the reactive forces. Spirits and men are both two stages in the transformations of smell. Men can

become spirits and spirits can become men, just as the individuals on the see-saw move from one position to another:
the one who is at the lower level rises to the higher position and
the individual who was at the higher level comes down to the
lower position.

Power and its acquisition is, however, based on the contact
between the powerful and the powerless. Since the two are
together and yet at the two ends of the metaphorical see-saw,
the Ongee cosmology can continue moving. It is this continuity
in life which the Ongees regard as a game of hide and seek. This
makes power transactable between the powerless and the
powerful through movement and contact; and this makes the
identity of the powerful and powerless dynamic, and capable
of shifts and switches, like the two individuals on the see-saw.

How the understanding of power relations and shifts of
identity derived from Ongee culture and praxis is of consequence is best seen in those instances where individuals active-
ly encounter the spirits by replicating the paths and patterns of
movement with which the spirits are associated. The initiate
and the *torale* are individuals who travel on the vertical axis of
movement within Ongee space, come into contact with the
spirits, and have distinctive powers because of that contact.

During the course of the initiation ritual, the novice's body
is prepared to ascend to the place of the spirits. As part of his
preparation, the novice must ingest only liquids and his body
must be 'heated' by massage. The ritual focus is on making the
novice like an infant, since there is a close identity between
infants and spirits, and on making the body light of weight and
emit smell. Discarding weight constitutes the essence of the
ritual of initiation. When the body of the novice is light and
emits a smell, it becomes particularly susceptible to being taken
away by the spirits. To ensure that the spirits come and take
away the novice, the Ongees offend them by hunting an excessive number of pigs in the forest. The time of the *tanageru*
ceremony is the period during which the spirits hunt in the
forest. Consequently, the hunting of pigs in the forest by the
Ongees during this time is regarded as *beti* ('a particularly
offensive and bad work'). The hanging of the leaf curtain is a
further means of ensuring that the spirits will come and take
away the novice.

The *torale*, in the course of his spirit communication, makes himself susceptible to being taken away by the spirits. In order to attract the spirits and induce a situation of spirit encounter (*enegeteebe*), the *torale* takes all implements, cutting edges, and ancestral bones that have been given to him by other Ongees to the forest. The *torale* exposes the bones to the winds in the forest and keeps the implements close by him so that he may carry them with him when the spirits come to take him away. The *torale*'s exposure of the bones and tools is an act of bodily submission to a situation in which the spirits are attracted to him and take him away. This is identical to the situation where the novice's body is processed and ritual activities are undertaken to attract the spirits to the novice and find him easy to take away. The novice's body is made light of weight; the spirit communicator lets the spirits draw out all the liquids from his body, making it light enough to be carried away. Thus both have to alter their body conditions to come into contact and move with the spirits. They are both seen by the community as being in extremely dangerous situations in which the individual involved is regarded to be as if dead.

The difference between the two individuals is that the *torale* goes on his own at the request of the community. The *naratak-wange* goes because of the efforts of the community as a whole, and he is also brought back to the community by the community. The *torale*'s visit to the spirits is an individual's effort and risk, especially in terms of being able to return. The *torale* may undertake the trip frequently within a seasonal cycle since there are only a few *torales*. Every time the *torale* goes up and comes back, the Ongees say that his body has to be taken care of in order to be ready to go up again.

Of the four individuals who were known to be practicing *torales* among the Ongees, the people reported that each of them had different powers based on the duration of their stays. Different *torales* 'went away with the spirits for many days or for a few days. A good *torale* stays with the spirits for many days, returns, and is soon ready to go again!' The *torale*'s journey, undertaken at the request of the community, is a distinctively practiced and acquired skill and power which benefits the entire community. The visit to the spirits undertaken by a novice in the course of the initiation is, however, a movement taken only once, after marriage, by a man in the course of his life.

The whole community is involved in the initiate's visit to the spirits. This aspect of movement up to the spirits by an individual through collective effort makes the *naratakwange*'s *enegeteebe* different from that of the *torale* who, as an individual, is responsible for all preparations and the final movement. The *torale* goes on his own and comes down alone because of his special capacity acquired through practice and the power gained from previous visits to the spirits. In other words, the experienced *torale* acquires the power to give to the spirits what they want from him, making his return possible. In the case of the initiate, the community is responsible for sending him up and is also responsible for preparing him to take steps so that he can return after negotiating with the spirits.

Although there are apparent differences between the individual effort of the *torale* and the collective effort needed for the initiate, some characteristic features remain common in the case of both the initiate and the spirit-communicator. Both the individuals lose body weight, both suffer an acute loss of smell, and, in the eyes of the Ongees, both the individuals become like spirits for a short duration, since a visitor to the spirits is regarded to be as if dead until he returns.

Both the *naratakwange* and the *torale* acquire a special position after they have undertaken their respective trips to the spirits. During the course of initiation, all males exhibit the power that the *torale* frequently shows to the community through his movement with and to the spirits. The Ongees regard the journey up to the spirits and the return journey as a means of endowing oneself with power which is described as 'a vision from above the forest'. This is a point of view which only the spirits have; men see this view only after they have died and have become spirits. As the Ongees put it:

Above the forest is a place from where you can see everything, the animals to hunt in various parts of the space, the things that are to be collected—things which have been lost, and the place where one should go to find things. It is there above the forest that you come to know where the spirits are going to be, where the winds will be going. It is only above the forest that man can see his wife as a widow—never on the island can you see your wife as your widow!

When the initiate and *torale* reach the place of the spirits, both negotiate with the spirits. Their negotiation involves giving away *maonole*, implements, and weapons to the spirits or promising that upon their return to the island they will provide the spirits with certain *maonole* and food. That the visiting Ongee can succeed in negotiating with the spirits is the most important aspect of the individual's power since negotiation is the only way an Ongee can ensure his return from the place of the spirits. The spirits allow the visiting initiate and *torale* to return only after they are assured that the return means future supply of what they want from the Ongees.

After giving the spirits the tools, the *torale* returns and shares with the rest of the community the information and knowledge he has gained from his contact with the spirits. The instructions provided by the *torale* make it possible for the entire Ongee community to undertake translocation, so that the Ongees may hunt without being hunted by the spirits. It is the *torale's* knowledge, gained by risking his life, undertaking a movement upwards, and becoming like the spirits, which enables him to sustain the balance between men and spirits.

The Ongees regard themselves as being in a relation of *gae?bebe* ('give and take') in a game of 'hide and seek'. It is the *torale* who has the power to conduct the game in accordance with the rules acceptable to both spirits and humans. The ability to make such rules pertaining to seasonal translocation needs the capacity and skill to move on the vertical axis and to communicate with the spirits. Every time the *torale* makes a trip to the place of the spirits he enriches his knowledge of Ongee space and acquires more power to navigate himself and the community through various dangerous and safe places which constitute space.

Since spirit encounters and coincidences of movement paths between men and spirits are avoided, continuity of life, i.e. birth of children within the community, becomes possible only when the situation of the spirits controlling the movements of men is changed. What the spirits do to men, in terms of making them move from place to place, men have to be able to do to the spirits. *Getankare gikonetorroka* is an example of how the Ongees manipulate the spirits. The Ongees offend the spirits by consuming all the honey. When the angry spirits

realize that all the honeycombs are being consumed by the Ongees, they become angry and leave the island. This is the inverse of the usual situations, i.e. in a specific season the Ongees leave a place when the place is visited by winds and spirits who come to the island to eat a particular food item. The Ongees expel the spirits from the island in order to gain all the honeycombs. Honeycombs are crucial to childbirth within the Ongee community. *Getankare gikonetorroka* exhibits the principle of power which Radcliffe-Brown (1964: 362-3) described as risking the anger of the spirits in order to obtain a desired end.

The hunting of pigs by the novice during the initiation ritual also demonstrates how angering the gods is a means of manipulation used by the Ongees in order to gain something. In this case the *naratakwange* is the negotiator–representative of the community, which has collectively offended the spirits in the course of the ritual. During his negotiations the novice, like the *torale*, promises to provide the spirits with what he has remembered as his *gilemame*. The novice's negotiation with the spirits also relies on invoking the identity between a child and a spirit, since the chosen *gilemame* are drawn from the products of the season in which the novice acquired his masticating powers and thus completed his transformation from a spirit to a human.

When the novice successfully returns, he is to be further processed and implanted with memory, i.e. the Ongees resocialize him. This resocialization process indicates that the novice undergoes a change in identity in the course of travelling to and from the residence of the spirits. The progressive changes that the novice undergoes begins with his being made light of weight as a child, then being carried with the spirits as if he were dead, and then being brought back as a child who is served food which needs no mastication. Finally, the novice's journey is completed when he is transformed into a fully matured individual male who is capable of hunting by himself in various parts and places of the island. The individual is only allowed to hunt all alone after his initiation, since his personal negotiation with the spirits makes him strong, safe, and powerful against any potential situation in which the spirits might take him away. The promise of providing *gilemame* to the spirits makes the novice an asset to the spirit community, since they will eat the promised *gilemame*. After marriage, all males go through *tanageru*. The Ongee men say:

> We all are concerned about the spirits who may take us away. We have the *gobolagnane*, but we also have made the *gilemame* available to the spirits since we started hunting alone in different places. Since we have provided *gilemame*, some spirits remember it so that when the spirits come to take us, we have some good spirits who do not want us to come up.

In other words, all individuals safely move and hunt and gather by using *gobolagnane*. The *gobolagnane* affect the movement of smell and of the spirits. It is the *gobolagnane* which help the Ongees to keep their horizontal axis of movement from coinciding with the vertical axis of spirit movement. By using the *gobolagnane*, the society succeeds in living and avoiding any contact with the spirits. However, the *torale* and the *naratakwange* are used by the society, as with the *gobolagnane*, to have contact with the spirits. Consequently, the result of these rituals has the same effect as that of the *gobolagnane*, i.e. the Ongees move without having an encounter with the spirits. Thus, the *torale*'s information gained by being above the forest and the *naratakwange*'s dealings with the spirits both have the same effect in terms of making Ongee movement safe and fruitful, as is the case with the *gobolagnane*.

The novice's return from the spirits, like the *torale*'s, benefits the entire community. In addition to the individual's acquisition of power to move safely from place to place in the course of hunting, the novice brings an end to the duration of *gengeyebe*, in which the winds and spirits are absent. The result of *getankare* and *tanageru* is that men succeed in impacting on the spirits. The end of the man-made season and the start of the spirits' seasonal cycle assures future childbirth and continuity of life. Deliberate and offensive activities, such as *getankare* and *tanageru*, show that by shifting identity, moving, and coming into contact with the spirits, the see-saw of Ongee cosmology and the game of hide and seek continues.

In the *chongojebe* domain the spirits are responsible for birth and death. However, during the processing and movement of the novice in the course of the temporal duration marked by *getankare* and *tanageru*, all the Ongees enact the birth and death and repeated birth of the novice. Therefore, the basic concern is not only for power with one powerful agent, but also the shifting of power from context to context, whereby the powerful can become powerless and vice versa. Ongees understand power and shifting identities in and through the movements on the two axes within space. In the context of the movement of winds, spirits, and smells, and of controlling all three by *gobolagnane*, the *torale* and the *naratakwange* are the most important, externalized, objectified, and reified elements in the Ongee world-view.

Power lies in the capacity to move and to make things move in relation to one's own movement. For example, both the *torale* and the novice first move towards the spirits, becoming more and more like the spirits since their bodies lose smell and weight. Consequently, the *torale* and the novice go through a kind of death and birth in their movement along the vertical axis. The birth and dying in the course of vertical movement is an aspect of the power which challenges birth and death, for which the spirits alone are responsible. This inversion of the relationship of life and death between man and spirit takes place in the course of an individual man's effort in moving along the vertical axis. The other outcome demonstrated in the novice's and *torale*'s movement is that the power they gain helps the rest of the community to move in relation to the spirits' movements into various places and in different seasonal durations.

Ongee cosmogonic myths give further support to the idea that the horizontal axis has to do with transformations and the vertical axis with power. One such myth concerns the preservation of fire by the monitor lizard and the civet cat during the great flood. Some of the ancestors escaped the flood by climbing up a tree. However, they left the greatest prize, fire, in the cooking pot. Monitor lizard, aided by his wife civet cat, grabbed the fire and took it up the tree. Ancestors who were unable to climb the tree became fish and other forms of marine life; those who could not descend from the tree after the flood became birds; and those ancestors who brought fire down from the tree when the water subsided became human beings. Monitor lizard and civet cat were the first man and woman.

Thus, the flood divided the ancestors into three forms of life confined to the three spatial divisions (sky, land, and water), forming the totality of species and of space in Andamanese taxonomy and cosmology. The ancestors who remained in the tree were in the world of 'light'; those unable to climb the tree entered the dark world of water; the ancestors who descended from the tree could walk across the forest and to the sea with power in the form of fire/smoke (the magical substance, i.e. *gobolagnane*, by means of which the natives move from land to sea forming the horizontal axis and transform other forms of life, such as pigs and turtles, into food).

The myth not only establishes the tripartite formation of space, but also distinguishes zoological groups on the basis of movement: those who fly in the world of light, those who walk in the world of shade, and those who swim in the dark world of water. Of all the animals formed by the flood, only the monitor lizard and the civet cat can move on both the vertical and the horizontal axes: by climbing up the tree they saved fire; after coming down they move on the horizontal axis using fire to transform other animal forms.

Since the vertical and horizontal axes map the totality of Andamanese space, anyone who can move on both axes has great power. Such is the case of monitor lizard and civet cat, whose movement is replicated by the *torale* and the initiate. The capacity of the monitor lizard and the civet cat to move along both axes not only maps the cosmology but makes explicit the notion of cartography in Andamanese cosmology—that power is the ability to move in the totality of space on both axes. Since the lizard and the cat are cartographers of both vertical and horizontal movement, they are, 'cosmosophically' speaking, the most powerful—the prime cartographers who map out the location of the power that can be acquired by movement. The lizard and the cat thus provide a navigational model to guide the native to power—as hunter, initiate, and spirit communicator. This map gives the native knowledge of where the spirits reside and how their power can affect him, representing both visual and physical topography and systematizing movement in relation to the invisible 'other world' as well.

Considering movement and its pattern as derived from ritual and myth, we get an idea of power and cosmology as related through the philosophy of movement. Power is always associated with the spatial coordinates, up and down—the vertical axis. The spirits come down, causing death; the lizard and the cat come down from the tree with the power of fire. The transformative effects of power are manifested on the horizontal axis. For example, the power of fire retained by the monitor lizard and the civet cat creates pigs confined to the forests, and turtles confined to the sea—spatial categories on the horizontal axis. Monitor lizard and civet cat with the fire could move and burn the animals of water and land. The scare of being burned kept the animals from attempting to move out of sea and forest. Magical substances are used by the natives to effect transformations during their movement on the horizontal axis, from powerlessness (in relation to the spirits) to safety. Otherwise, everyone in horizontal movement is powerless, subject to the spirits moving on the vertical axis.

Within the Ongee culture, movement is a metaphor through which power, danger, and safety are not only expressed ideas but become objects which are manipulated by the power-seeking Andamanese in any form of movement. When the Andamanese move, their notions about time and space fuse, and thus enable them to construct their cosmology. This is evident from certain *gobolagnane* being essential for movement in any form. *Gobolagnane* make all the movements through space constant, i.e. safe from the influence of the spirits. For the Ongees it is the spirits who make different temporal experiences in space.

Radcliffe-Brown (1964) describes these essential objects, the *gobolagnane*, as socially valuable and/or magically potent. They are essential since they provide protection against the ubiquitous spirits. In the initiation ceremony, the initiate gains power by undertaking a trip to the spirits and coming down. It is this received power which generates movement and makes the novice a hunter. Without it, he himself becomes the hunted. This formula has a syllogistic implication: power generates movement, and to move power is required.

Since without power to move, a person will be hunted by the spirits, there is a coincidence created in Ongee ritual. Both the *torale* and novice attract and submit to spirits. In other words, the two individuals (hunters) change into the hunted in order to gain power. When the Ongee hunter becomes the hunted, he receives the power to hunt again. The Ongees say, 'We kill so that we live to kill more'. Thus, to gain power, the Ongee switches identities, as with *malabuka*.

Another Andamanese myth taken from Man's (1885) account explains the idea of shifting power and switching identities. In ancient times, fire was under the exclusive control of Biliku (also called Dare), who gave the fire to Sir Prawn. The kingfisher stole the fire to roast; he ate and then went to sleep. Then the dove stole the fire and gave it to the Andamanese, who were taught by Biliku to cook their food. Kingfisher's theft was punished when Biliku cut off the bird's head and transformed his status from that of ancestor to mere bird. Thereafter, the kingfisher could only eat raw food; the Ongees could cook but lost the ability to fly. Fire then transforms raw food to cooked food, and distinguishes birds from men. It should also be noted that the Ongees consider roasting more dangerous than boiling; the roasted food served on ritual occasions marks exocuisine; the Ongees otherwise prefer boiled food, endocuisine.[1] Thus, roasted food is natural and boiled is cultural, and both are distinguished by the use of fire. Fire also distinguishes, by its possession, animals and humans: those ancestors who were afraid of or were burnt by fire became birds, beasts, and fish—cut off from the human society which 'from that moment constitutes itself around the fire' (Radcliffe-Brown, 1964: 342).

Kingfisher's stolen fire enabled him to cook; but once the fire was lost, he had to eat raw food and assume a natural identity. Fire, then, differentiates men from animals, nature from culture; its absence destroys the nature/culture distinction. Ongee mythology thereby presents events in which the acquiring of fire, a power, creates a distinction between men (culture) and animals (nature). The absence of this power again switches nature and culture: men without fire are like animals, and vice versa. Kingfisher exemplifies this mythologic; by stealing fire he is like men, but having lost its use he is just a bird. The implication is that possession of power constitutes

[1] Andamanese, including Ongees, boil the meat and serve it. Sometimes, when a large number of neighbours visit a camp-site the meat served is pit roasted. Visitors are rarely served boiled meat.

one as a cultural entity, which again presupposes that to gain power one must start as a natural entity.

Assuming this scheme, let us consider the significance of the kingfisher: the fire stealer as revealer. He has an indexical value, as his bright red plumage reminds us of the stolen fire; social value in the incident of his theft and subsequent punishment—which explicates the nature/culture distinction in his relation to his descendants, men. But above all is the kingfisher's capacity to generate distinctions between nature and culture so that in the mythological event nature and culture coincide. Thus mythical events order nature for individuals.[2]

According to Radcliffe-Brown, the power of Andamanese ritual lies in its conveying to the individual his dependence on society and upon its social valuables, such as fire. Like the kingfisher, the ritual participant gains power from the process of coincidence in the mythical event—the switching from cultural to natural and back again within his own social time. (For instance the ritual of *tanageru* and the myth of the first *tanageru*.) Such power is seen in the kingfisher's fusion of man-like cultural identity (eating cooked food) with natural (still being a bird), and the subsequent loss of his ancestral identity. This conversion of the nature/culture duality into a unity by switching between the two is one way power is generated in Ongee rituals. This sort of formulation goes beyond Radcliffe-Brown's notion of Ongee ritual as a meta-social form of projective system capable of converting social sentiments and values into communication.

The notion that rituals generate power through switching natural and cultural identities can be traced to the works of Hocart. According to Hocart, one thing is made equivalent to another—which we understand as symbolic. 'If you cannot act on A by acting on B there can be no ritual' (Hocart, 1970: 45). Ritual participants seek to establish 'an identity between man and the ritual objects, between ritual objects and the world, a kind of creative syllogism' (ibid.: 64). In other words, there is a classification of nature and culture. Men who cook become cultural through association with fire, and animals who do not

[2] Cf. the Ongee initiate's activity in the course of initiation and the myth of the first *tanageru*.

cook become natural through disassociation from fire. These equivalences established among various things that may appear to be quite dissimilar to an outsider, however, affect the one member of a class by acting on another.

I have considered movement and its patterns derived from rituals and myths to understand Ongee cosmology and notions of power. Power is always associated with the spatial coordinates of up and down, that is, the vertical axis. For example, the spirits come down and cause death. The transformative effect of power is manifested on the horizontal axis, where Ongees hunt in the sea or at the forest without getting hunted by the spirits.

Gobolagnane are used by the Ongees to bring about transformations during the course of movement on the horizontal axis. It is this use of *gobolagnane* which makes Ongees safe in relation to powerful Ongee-hunting spirits and makes hunted animals powerless in relation to the Ongees.

Movement is a metaphor through which power, danger, and safety are not only expressed ideas but become objects which are manipulated by the power-seeking Andamanese in every form of movement.

Both the *torale* and the initiate as individual hunters become the hunted in order to gain power. When the Ongee hunter becomes the hunted he receives the power to hunt again.

Being powerless is a stage before being powerful: the power to dominate originates in submission; deliberate coincidences enable future conjunctions to be avoided; and from death emerges birth.

All these ideas come together in the Ongee proverb:

Eneykutata gateyeybeh — Enekutata benchame ma Konyune rotanka kut gateyebeh benchamema (We kill so that we live to kill more and live more!)

In my consideration of the Andamanese ethnographic material, I have argued that Ongee cosmology is closely related to movement, a relationship that occurs frequently in Andamanese myth and ritual. Ongee preparation prior to movement indicates an awareness of space, of the spirits residing in it, and of the prospective dangers—in relation to cosmology as a map of power. The map, in turn, is a medium through which Ongee ideology is articulated in terms of movement. Movements in relation to space are ritually ordered and require the generation of power. Further, movement also require *gobolagnane*, magical substances. Thus, we find a variety of attempts to map out positions of power in myths, rituals, and experiences of transformation in day-to-day life where there is a concern for 'above the forest' while living in the forest.

Glossary

Aa	Bow.
Aateelebey	To talk.
Akwabeybeti	Doing something which causes the winds and spirits to leave or move away from a place; bad or offensive work.
Akwanegenegabe	Reincarnation; one who dies is born again.
Alame	Red clay paint.
Alankare	Singing done in the presence of pregnant women inviting the child to emerge from the womb.
Ale	Child.
Amboro	Heavy; state of well-being.
Ame	Bud of plant.
Andangale	Cooking area at time of the ritual *tanageru* where those who serve and those who consume food come together.
Angachee	Wife.
Ayuge	Monitor lizard.
Batitujuney	Horizon: place where land, sea, and sky meet.
Benchamee	Dead, death, gone forever; will not revive; being taken away (upwards) by spirits; insult.
Berale	Circular large shelter with conical roof built to accommodate more than one nuclear family and a place where ancestral bones are stored.
Beti	Offensive; insult; transgression.

Be?yeche	Cooking fire.
Bonee	Light; illumination; fire.
Bulledange	Jackfruit.
Cha?ntembobey	Ants.
Chendange	Seeds of the creeper belonging to the family of *Ipomoea pesceprai*. When it is thrown into a fire it makes a loud explosion that drives spirits away.
Chenekwa	Arrow; one that causes flow of smell.
Choge	Food; fish.
Chongojebe	Situation or phase in which spirits act and humans receive impact of those acts.
Daboja	Fruit of *Bruguiera gymnorhiza* (Rhizophoraceae).
Dange	Living things; things that can be cut, shaped, or bound—such as the human body, wood, plants, and canoe.
Dare	Seasonal duration associated with winds from the south-east that bring the spirit Dare to the island (May to July); a spirit associated with the south-east winds.
Detababe	Acquiring weight or heaviness by individuals involved in a ritual; weight discarded in the process of a ritual; impact aspect of ritual. One who is ritualized loses weight and those involved in the ritual gain weight.
Dobolobolobe	Pattern in which a snake moves; curved and wavy lines; pattern of movement of the spirits.
Dolakambey	Small; light; thin.
Durru	Thunder; noise made by pulling a bowstring; noise made by the monitor lizard.
Eahambelakwe	Turtle hunters.
Eahansakwe	Pig hunters.
Ebebarro	Knowledge; sight from above.
Ejemotto	Intrinsic quality of *toleudu*: to cool down the body.

Ekullukutta	Intrinsic quality of *alame*: to heat up the body.
Ekwacele	Path; path connecting the island with the spirits' residence; path over which a child travels from the spirits' residence to a woman's womb; path which a dead person travels to spirits' residence.
Elokolake	Immobile; one that cannot move on its own.
Enakyu?la	Power; capacity to replace or displace.
Enechekebe	Cartographic knowledge, map; knowing how and where to move and in which direction.
Eneedabatanebe	Isolation.
Enegeteebe	Embrace; encounter; a meeting with spirits that results in the spirits taking a human body to their residence.
Enelukwebe	Painted designs on the body.
Eneratetangeyabe	Hunting pigs during *tanageru*.
Enerengewa	Non-living things without bones; absorbers of smell.
Enetebe	Painted designs on the face.
Eneteea	Body internal that exits out of the body and moves around while an individual sleeps or while an individual is in contact with the spirits.
Eneyabegi	Woman's milky breast; mythical ancestral figure.
Eneyachuge	Swings.
Eneyagegi	Woman's grass apron; mythical ancestral figure.
Eranabeti	Anger; upset; disturbed.
Erotakabe	Food to be avoided.
Gae?bebe	Giving and taking relations with spirits.
Gakhwe?kabe	Reptiles.
Galawelatetaye	Interdependent.
Galujebe	Scattering pig's blood after killing it in the forest.

Gamyebe	High tide; massaging the body in an upward direction.
Gananbube	Cutting pigs killed in the forest without spilling or scattering blood.
Gandema	Brave.
Gatee	Skin.
Gawakobe	To talk with spirits, children, or ancestors.
Gayekwabe	Smell tied tightly; hunter of animals.
Gebo	Flesh; meat.
Gebokwela	Living things with bones; capacity to emit smell.
Geduba	Experienced symptom of lower half of body being heavy.
Geerange	Bones.
Geeroyebe	Horizontal movement.
Gegamebe	Something that can float; one who has memory.
Gegi	Tubers.
Gekalakwebe	To make something happen; one who acts; one who controls.
Gemey?be	Manipulation of body weight and condition by massaging.
Genekula	Growth; process of smell and liquids becoming solid.
Gengegetebey	Spirits 'stealing' from men.
Gengeyebe	Phase or duration in which human beings act and the spirits receive the impact of those acts.
Gen?yochaye	Smoke.
Getankare	Ritual emptying Ongee space of winds and spirits.
Gigabawe	Singing; causing attraction.
Gikonetorroka	Ritual; to discard body weight.
Gilemame	Proscription and prescription pertaining to consumption and production.
Gingetigye	Cyclones.
Girorobuke	Gift.

Gitekwatebe	Release of smell; flow of death; hunting an animal; one who is hunted.
Gobodegalemba	Season.
Gobolagnane	To go and bring, valuables; things which effect smell and the capacity to move; collective term for clay paints, fire, and bone ornaments.
Gobomamey	Full moon, night of highest tide; dangerous bleeding.
Goke?	Initiators making the initiate light.
Gucheyakolabe	Walking behind a person and stepping on the track made by the person walking ahead.
Gugekwene	One who receives the impact of another's action; one who is controlled and effected by others.
Gukwe	To seek; to search.
Gukwelonone	Hide and seek.
Gwe?yekala	Togetherness and interdependence; things that can move any- and everywhere; quality of permeability like the spirits and winds possess; one's immediate family.
Gyambabe	Danger.
Gy?ole	Light; state of being unwell.
G?enabe	Massaging the body in a downward direction; pushing and pressing down the body's liquids.
Ibee	Bad smell, stink; danger caused by decay.
Ibeedange	Lower jawbone.
Idankuttu	Big; heavy; fat.
Igagame	The change brought about in temperature and moisture because of the movement of winds, spirits, and human beings from one place to another.
Ijababe	Cutting.
Ikatallabe	Space made or turned dry through loss of liquids.
Ikulukutta	Sun's fire; quality of heat.

Inachekame	To forget.
Injube	Space.
Kalakala	Binding; cooking area where eaters come to eat.
Kamakulehlekwe	Efficacy; intrinsic quality of *gobolagnane* to transfer their virtues.
Kame	Scaffold; sleeping platform.
Kayare	Low tide; form of body massage in which body weight is pushed down.
Kekele	Civet cat.
Ketukabe	Copulation; something which is comforting.
Ketula	Phallus; digging stick.
Ketulabe	Carved wooden phallus.
Koney	Lightness.
Keye?	Yellowish skin derived from the stems of *Dendrobium* species of orchid, has the capacity to counter with heat-generating substances.
Korale	Lean-to; shelter; family unit.
Kuge	Stone; anger.
Kugebe	War.
Kwalakangne?	Season of south-west winds when spirit Kwalakangne? resides on the island (August–October); spirit associated with south-west winds.
Kwaya	'Presence of either this or that.'
Kwayabe	Smell.
Kwayaye	Smelling; smell emission; tides.
Lemolakeye?e	Silence.
Lololobe	Shivering cold.
Lololokobe	Earthquake.
Lonone	To hide; to cover up.
Ma	'No'; negative marker.
Macekwe	Ancestors.
Makwekatakokowebe	Successful hunting.

Malabuka	Coincidence; unexpected intersection of paths of movement.
Maonole	Tools.
Matee	Body external.
Mayakangne?	Season duration when winds from the north-east bring the spirit Mayakangne? to the island (October–February); spirit associated with the north-east winds.
Megeyabarrota	Section of the forest associated with one's mother's group and/or wife's group.
Mekange?getelakwe	Bad luck; misfortune, not finding something; unexpected happening; a negative spirit encounter.
Makwekatakokowebe	Good luck; good fortune; expected or anticipated happening; finding something; positive spirit encounter.
Melame	One who walks in front, knows the way, and leads the way; having navigational skills.
Menyakuttu	Affection.
Metakabe	*Torale*'s flight upwards along with the spirits.
Mijejeley	Friend.
Mineyalange	Communication between human beings and spirits (through application of clay paints to the human body); to remind; to remember.
Mobetega	Young and ignorant.
Monatandunamey	'My hunger'; entire seasonal cycle.
Mutarandee	Mother's brother.
Muteejeye	Initiator.
Nakwarabe	Unavailability of honey because of the dry season.
Nanchugey	Place.
Nanguchumemy?	Process of solid becoming liquid and finally releasing smell; process of decay.
Naralanka	Turtle.
Naratakwange	Chambered nautilus shell; novice.

Nateelabe	Inviting someone to see something.
Nuwegerro	Red clay.
Nya?nya	Shellfish; shrimps; lobsters.
Oame	Fingers.
Obeikwele	Little finger.
Obonaley	Residence for all the bachelors; home where all men collect.
Obotabe	Thumb.
Oikanare	Spit.
Olo	Experienced symptom of top half of the body being light.
Onolabe	Dance.
Onotangile	Sweat.
Otenduabe	Attraction; an effect of singing.
Ougeru	Red, like burning wood.
Talabuka	Conjunction; desired intersection of axis movement.
Tamale	To make; to bring together; to build; to fabricate.
Tambanua	Pig.
Tana	Blue like ash; body of spirits.
Tanageru	Ritual of initiation of young men.
Tananey	Sending signals by beating the buttress of a tree.
Tanja	Honey.
Tea	Shape or form of living things.
Tegule	Water spout.
Tejage	Sharpening stones; cutting edges.
Togai?kwanenge	Trick.
Tolakebe	To cut apart; to split open.
Toleudu	Yellow and white clay paint.
Tombowage	Cicada grubbies.
Tomya	Spirits.
Tonkuloo	Sun.
Totaley	Activity; act; work.
Totoaate?	Far away places where only winds can go.

Tototey	Winds.
Torale	Seasonal duration from March to April with no winds experienced; season associated with collection of honey; a person who communicates and moves with spirits; to extract.
Tugey	Bird.
Tukuree	Fire.
Tukwengalako	Tree *Dipterocarpus icanus Rosib.*
Ulaateye	Body pain caused by loss of body liquids; decrease in weight and the experience of heat by the body.
Ule ule	Again and again.
Ulokwobe	Binding and weaving.
Umma	Tube with a design woven on it to enable a novice to consume all the liquid food served to him during the *tanageru* ritual.
Utokwobe	A spirit coming down and changing into a foetus to emerge from a woman's womb; childbirth.
Uwe	Clay.
Uwekella	White clay.
Uwekokey?	Clay paints.
Uye?uye	Stealing; taking without being noticed; inappropriate appropriation.
Wabe	Camp-ground surrounded by *korale*.
Wabekomabe	Vertical movement.
Yeenehye?bagabeh	To continue cutting and tying; to make it possible to live; life.
Yenagoranka	Ornaments made from a dead person's bones.

Bibliography

Agarwal, H. N. 'Reproductive Life of an Ongee Woman', *Vanyajati* 15 (1967): 139-49.

Basu, D. N. 'An Account of Andamanese Dance, Song and Mythology', *Indian Folklore* 2 (1957): 91-6.

—— 'The Present Day Andamanese Culture', *Indian Folklore* 2 (1959): 20-4.

Bloch, J. 'Prefixes et Suffixes en Andaman', *Bulletin de la Société de Linguistique* 45 (1949): 1-46.

Bose, S. 'Economy of the Ongee of Little Andaman', *Man in India* 44 (1964): 298-310.

Bourdieu, P. *Outline of a Theory of Practice* (Cambridge: Cambridge University Press, 1977).

Cipriani, L. 'Report on a Survey of Little Andaman 1951-52', *Bulletin of the Department of Anthropology* (Calcutta) 1 (1954): 61-82.

—— 'The Opening of the Little Andamans', *Bulletin of the Department of Anthropology* (Calcutta) 2 (1954):45-55.

—— 'A Survey of Little Andamans during 1954', *Bulletin of the Department of Anthropology* (Calcutta) 3 (1954):66-94.

—— 'Hygiene and Medical Practice among the Ongees', *Anthropos* 56 (1961):481-500.

—— 'Recent Anthropological Work in Little Andamans', *Current Anthropology* 3 (April 1962): 208-14.

—— *The Andaman Islanders*, edited and translated by D. Tylor Cox (New York: Praeger, 1966).

Codrington, R. H. 'Mana', in *Reader on Comparative Religion*, edited by W. Lessa and E. Z. Vogt (New York: Harper, 1965).

Colebrook, Lt. R. H. 'On the Andaman Islands', *Asiatik Research* 4 (1795): 385-95.

Corbin, A. *The Foul and the Fragrant* (Cambridge: Harvard University Press, 1986).

De Roepstorff, Fr. Ad. *Nicobar and Andaman Isles* (Calcutta: Office of the Superintendent of Government Printing, 1875).

Dumont, L. *Homo Hierarchicus* (Chicago: University of Chicago Press, 1970).

Dumont, L. 'The Anthropological Community and Ideology', *Social Science Information* 18 (1979): 785-817.

—— 'On Value' (Radcliffe-Brown Lecture), *Proceedings of the British Academy* 66 (1980): 207-41.

Durkheim, E. *Elementary Forms of Religious Life* (New York: First Free Press, 1965 (1915)).

Durkheim, E., and M. Mauss. *Primitive Classification* (London: Cohen and West, 1963).

Endicott, K. M. *An Analysis of Malay Magic* (Oxford: Oxford University Press, 1970).

—— *Batek Negrito Religion* (Oxford: Clarendon Press, 1979).

Firth, R. *Tikopia, Ritual and Belief* (Boston: Beacon Press, 1967).

Fortes, M. 'Time and Social Structure: An Ashanti Case Study', in *Social Structure: Studies Presented to A. R. Radcliffe-Brown* (Oxford: 1949).

Ganguly, P. 'Religious Beliefs of the Negritos of Little Andaman', *Eastern Anthropologist* 14 (1961): 243-8.

Ganguly, P., and A. Pal Ganguly. 'Ongee Harpoon and Spear', *Anthropos* 58 (1963):557-60.

Geertz, C. *Negara, The Theatre State in 19th Century Bali* (Princeton: Princeton University Press, 1980).

Gell, A. 'Magic, Perfume and Dream', in *Symbols and Sentiments*, edited by I. Lewis (New York: Academic Press, 1977).

Harrer, H. *Die Letzten Funfhundert: Expedition zuden zwergvolkern auf den Andemanen* (Frankfurt: Main Ullstein, 1977).

Harrison, J.E. *Themis: A Study of Social Origins of the Greek Religion*, 2nd rev. ed. (Cambridge: Cambridge University Press, 1937 (1912)).

—— *Ancient Art and Ritual* (New York: Oxford University Press, 1913).

Hocart, A. M. 'Mana', *Journal of the Royal Anthropological Association* 14 (1914): 97-101.

—— *Kings and Councillors* (Chicago: University of Chicago Press, 1970).

Howell, S. Some Issues in the Study of Malay Aboriginal Ethnography', M. Litt. thesis, Oxford University, 1977.

—— *Society and Cosmos* (Singapore: Oxford University Press, 1984).

Inden, R. 'Orientalist Construction of India', *Modern Asian Studies* 20 (1986): 401-46.

Kochar, V. K. 'Ongee Dance', *Folklore* 6 (1965): 350-7.

—— 'A Note on Some Ongee Drawings', *Bulletin of Cultural Research Insititute* (Calcutta, 1965).

Kloss, C. B. *In the Andaman and Nicobars* (Delhi: Vivek Publications, 1971 (1903)).

Langham, I. *The Building of British Social Anthropology* (Holland: D. Reidel Publishing Company, 1981).

Leach, E. *Political Systems of Highland Burma* (Boston: Beacon Press, 1967 (1954)).

—— 'Kimil: A Category of Andamanese Thought', in *Structural Analysis of Oral Tradition*, edited by P. Maranda *et al.* (University Park: University of Pennsylvania Press, 1971).

Lesa, W., and E. Z. Vogt. 'Mana and Taboo', in *Reader in Comparative Religion*, edited by W. Lesa and E. Z. Vogt (New York: Harper Books, 1965).

Lévi-Strauss, C. *Structural Anthropology* (New York: Basic Books, 1963).

—— *The Naked Man* (New York: Harper & Row Publishers, 1981).

Malinowski, B. *A Scientific Theory of Culture and Other Essays.* (Chapel Hill: University of North Carolina, 1944).

—— *The Dynamics of Culture Change* (New Haven: Yale University Press, 1945).

Man, E. H. *On the Aboriginal Inhabitants of the Andaman Islands* (London: Royal Anthropological Institute, 1885).

Mann, R. S. *The Bay Islanders* (Bidisa, Bihar: Institute of Social Research and Applied Anthropology, 1979).

—— 'Animism, Economy, Ecology among Negrito Hunters and Gatherers', in *Nature, Man-Spirit Complex in Tribal India*, edited by R. S. Mann (New Delhi: Concept Publishing Co., 1981).

Mathur, L. P. *History of Andaman Nicobar Islands 1756-1966* (Delhi: Sterling Publications, 1968).

Mauss, M. *Manuel d'Éthnographie* (Paris: 1947).

—— *The Gift* (London: Cohen & West, 1954).

Mouat, F. J. *The Andaman Islanders* (Delhi: Mittal Publications, 1979 (1863)).

Munn, N. 'Symbolism in a Ritual Context: Aspects of Symbolic Action', in *Handbook of Social and Cultural Anthropology*, edited by John J. Honigmann (Chicago: Rand McNally College Publications, 1973).

Parsons, T. *Essays in Sociological Theory: Pure and Applied* (Cambridge: Harvard University Press, 1949).

Parsons, T., and E. Shils (eds.). *Toward a General Theory of Action* (Cambridge: Harvard University Press, 1951).

Portman, M. V. 'Disposal of the Dead among the Andamanese', *Indian Antiquary* 25 (1896): 56-7.

—— 'The Andamanese Fire Legend', *Indian Antiquary* 26 (1898): 14-18.

—— *History of Our Relations with the Andamanese* (Calcutta: Government Printing Office, 1899).

Radcliffe-Brown, A. R. 'Notes on the Languages of the Andaman Islands', *Anthropos* 14 (1914): 36-52.

—— *The Andaman Islanders* (New York: Free Press, 1964 (1922)).

Roy, B. C., and P. Ganguly. 'Some Ceremonial Customs in Ongee Life Cycle', *Folklore* 2 (1961): 368-74.

Sarkar, S. S. 'Ongee Population and Settlement', *Anthropos* 55 (1960): 561-3.

Schmidt, W. 'Riluga: The Supreme Being of the Andamanese', *Journal of the Royal Academy of Anthropology* 10 (1910): 66-72.

Schneider, D. M. *American Kinship: A Cultural Account* (Englewood Cliffs, NJ : Prentice-Hall, 1968).

Service, E. R. *The Hunters* (Englewood Cliffs, NJ: Prentice-Hall, 1966).

Singh, I. N. *The Andaman Story* (New Delhi: Vikas Publications, 1978).

Smith, R. *The Religion of the Semites* (New York: Meridian Books, 1956).

Sperber, D. *Rethinking Symbolism* (New York: Cambridge University Press, 1975).

Tambiah, S. J. 'World Conqueror and World Renouncer', *Cambridge Studies in Social Anthropology*, no. 15, Cambridge University Press, 1976.

—— 'A Performative Approach to Ritual' (Radcliffe-Brown Lecture), *Proceedings of the British Academy* 65 (1979).

Temple, R. C. *1901 Census Report of India*, vol. 13 (Calcutta: 1903).

—— *Remarks on the Andaman Islanders and Their Country* (Bombay: British India Press, 1930).

Thompson, A. 'Description of Andamanese Bone Necklace', *Journal of the Royal Anthropological Institute* 2 (1881): 295-309.

Tikader, B. K., and A. K. Das. *Glimpses of Animal Life of Andaman Nicobar Islands* (Calcutta: Zoological Survey of India, 1985).

Tuan, Yi-Fu. *Space and Place* (Minneapolis: University of Minnesota Press, 1977).

Turner, V. *Forest of Symbols* (Ithaca, NY: Cornell University Press, 1967).

—— 'Introduction', in *Forms of Symbolic Action*, edited by Robert E. Spencer (Seattle: University of Washington Press, 1969).

Van-Gennep, A. *The Rite of Passage* (London: R.K. Paul, 1960).

Weber, M. In *The Theory of Social and Economic Organization*, edited by T. Parsons (Glencoe, Illinois: Free Press, 1957).

Index

act–impact, relationship of, 168
adornment, body, 254–5, 267
AAJVS (Andaman Adim Jan Jati Vikas Samiti), 6, 13–14
anger: between spirits and men, 37–44; caused by offensive acts in myth, 7–8; expressed by men in myth, 10; expressed by spirits in myth, 102; expressed in ritual, 271, 272–5, 291
ants, 31, 124. *See also* paint

binding. *See* tying
birds: clan identity with, 21, 127; movement of, 63, 65, 107; relationship with spirits, 124
birth, 19–20, 80–2, 183
blood: menstrual, 121, 185–6; use of in *tanageru*, 196, 221, 223, 232, 247
body: effected on during *tanageru*, 253–4, 277; homology with nautilus shell, 227–9; internal and external, 98, 153, 214, 246; memory, 265; rituals affect-

ing, 164–5, 171, 203–4, 216–17; use of *gobolagnane*, 280. *See also* paint
bones, 137, 144. *See also* *gobolagnane*
Bourdieu, Pierre, 72
British, administration of Andamans, 4–6

cat, civet, 9, 31–2, 63, 65–6, 180
child, 242, 259, 278; conception of, 183–9
chogele 233–6
chongojebe, 168–9, 203, 207
choreography, 180, 253
coconut, 11, 13
colour, 204–5
construct, cultural, 71
cooking, 268
cosmology, 52–3, 60–2, 72–3, 79, 83, 113
counting, 73
crab, 32, 65, 276–7
crying, 228
cutting, 86, 112, 131, 143, 156, 223

dance, 10, 180, 256–8
dange (living things), 94, 96, 138–9

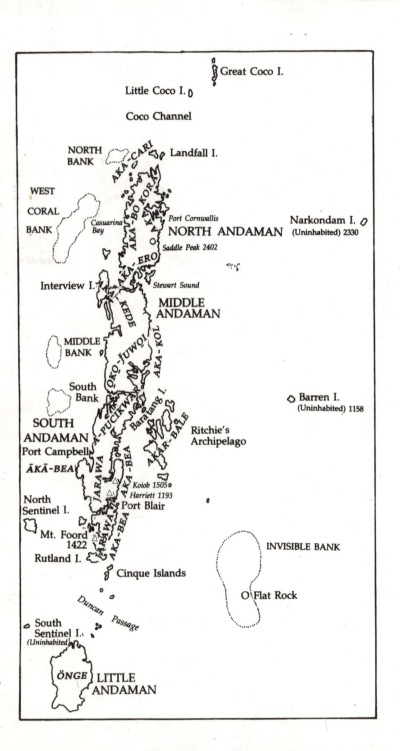

Great Coco I.

Little Coco I.

Coco Channel

NORTH
BANK

AKA-CARI

Landfall I.

WEST
CORAL
BANK

Casuarina
Bay

AKA-BO KORA

Port Cornwallis

NORTH ANDAMAN

Narkondam I.
(Uninhabited) 2330

Saddle Peak 2402

AKA-ERO

T-AKA-YERE

Interview I.

Stewart Sound

MIDDLE
ANDAMAN

KEDE

MIDDLE
BANK

OKO-JUWOI

AKA-KOL

South
Bank

Barren I.
(Uninhabited) 1158

SOUTH
ANDAMAN

A-PUCIKWAR

Baratang I.

Port Campbell

AKAR-BALE

Ritchie's
Archipelago

ĀKĀ-BEA

JARAWA

AKA-BEA

AKA-BEA

Koiob 1505

Harriett 1193

North
Sentinel I.

Port Blair

Mt. Foord
1422

ARAWA

AKA-BEA

Rutland I.

INVISIBLE BANK

Cinque Islands

Duncan Passage

Flat Rock

South
Sentinel I.
(Uninhabited)

ÖNGE

LITTLE
ANDAMAN